D0872917

Useful Bodies

PUBLISHING FOR THE WORLD
125 Years
THE JOHNS HOPKINS UNIVERSITY PRESS

Useful Bodies

Humans in the Service of Medical Science
in the Twentieth Century

Edited by Jordan Goodman,
Anthony McElligott, and Lara Marks

The Johns Hopkins University Press
Baltimore and London

The Johns Hopkins University Press
2715 North Charles Street
Baltimore, Maryland 21218-4363
www.press.jhu.edu

Library of Congress Cataloging-in-Publication Data

Useful bodies : humans in the service of medical science in the
twentieth century / edited by Jordan Goodman, Anthony
McElligott, and Lara Marks.
     p.   cm.
Includes bibliographical references.
   ISBN 0-8018-7342-8 (hardcover : alk. paper)
   1. Human experimentation in medicine—History—20th
century. I. Goodman, Jordan. II. McElligott, Anthony,
1955– III. Marks, Lara, 1963–
   R853.H8 U846 2003
   619'.98'0904—DC21   2002152398

A catalog record for this book is available from the
British Library.

Earlier versions of the chapters in this volume were first given
as contributions to a workshop on human experimentation
held at the Wellcome Institute for the History of Medicine in
London on 3–4 September 1998.

Contents

Acknowledgments   vii

1   Making Human Bodies Useful: Historicizing Medical
Experiments in the Twentieth Century   1
Jordan Goodman, Anthony McElligott, and Lara Marks

PART I: What Is a Human Experiment?

2   Using the Population Body to Protect the National Body:
Germ Warfare Tests in the United Kingdom after
World War II   27
Brian Balmer

3   Whose Body? Which Disease? Studying Malaria
while Treating Neurosyphilis   53
Margaret Humphreys

PART II: Who Experiments?

4   Human Radiation Experiments and the Formation of
Medical Physics at the University of California,
San Francisco and Berkeley, 1937–1962   81
David S. Jones and Robert L. Martensen

5   "I Have Been on Tenterhooks": Wartime Medical
Research Council Jaundice Committee Experiments   109
Jenny Stanton

6   See an Atomic Blast and Spread the Word: Indoctrination
at Ground Zero   133
Glenn Mitchell

PART III: Whose Body?

7   Injecting Comatose Patients with Uranium: America's
     Overlapping Wars Against Communism and Cancer
     in the 1950s   165
     Gilbert Whittemore and Miriam Boleyn-Fitzgerald

8   Writing Willowbrook, Reading Willowbrook:
     The Recounting of a Medical Experiment   190
     Joel D. Howell and Rodney A. Hayward

     List of Contributors   215

Acknowledgments

We would like to thank all of the participants, speakers, and audience for a very successful meeting. Tilli Tansey was very enthusiastic about the project and gave it her full support. We would like to thank her for that. The editors are grateful to the authors and to Jacqueline Wehmueller of the Johns Hopkins University Press for their patience and cooperation in bringing this project to a conclusion. We would also like to thank the Wellcome Trust for their financial support and Wendy Kutner for her unstinting administrative assistance.

Useful Bodies

CHAPTER ONE

Making Human Bodies Useful

Historicizing Medical Experiments in the
Twentieth Century

Jordan Goodman, Anthony McElligott, and Lara Marks

Human experimentation has its historians but not its history. In a pioneering attempt to outline how a history of human experimentation might be undertaken, William Bynum in 1988 pointed out that the history of this very important topic has not been explored.[1] Fifteen years later, little has changed. As a start in the project of historicizing human experimentations, of which this book is a part, we propose a rough typology based, not around the familiar doctor-patient or scientist-subject axis, nor on Bynum's valuable types-of-medicine approach, but rather on the role of the state as actor, legitimator, and provider.[2]

Why are we focusing on the state and not on the experiments per se? The simple reason is that the relationship between science and its subjects is not easy to historicize because its empirical disclosures come packaged as case studies and these, as we argue, are difficult to arrange and structure to give historical insight. Focusing on the nature and degree of the involvement of the state, on the other hand, provides the kind of sharp tool that unlocks the context and the practice of human experimentation as a historical process. We are, therefore, in total agreement with Gert Brieger

when he stated that a "mere catalog of human experiments, while interesting and perhaps even instructive, is not sufficient."[3]

Hence, we suggest a rough periodization relating to the involvement,
both direct and indirect, of the state: pre-state (before the 1930s); state
(1930s to 1960s); and post-state (1960s onward). Of course, we are not suggesting that these are rigid boundaries. The periodization is more fluid
than at first appears; boundaries are soft rather than hard, continuous
rather than discrete. It is, in short, more of a heuristic device than a fixed
regime.

Most historians would agree that the use of human subjects in nontherapeutic experimentation is a relatively recent phenomenon and that
the emergence and rise of this practice coincided with a more general scientization of clinical medicine toward the end of the nineteenth century
in both Europe and the United States. As such, human experimentation
is part of a process in the history of medical experimentation, or knowledge making, which also included the increasing use of animals in experiments and the decreasing practice of self-experimentation.[4]

For most of the nineteenth century, as David Rothman has argued,
"human experimentation was a cottage industry, with individual physicians
trying out one or other remedy on neighbors or relatives or themselves."[5]
By the end of that century, however, this began to change as the boundaries of scientific knowledge were pushed back. Along with the exploration
of the earth's "dark continents," the human body itself had become the subject of exploration—and conquest. But the site of individual experimentation then gave way to more general terrain that took in society per se.
Thus, while the nineteenth century sought scientifically to release the
"truths" of the inner self (and here one only need refer to Dr. Jekyll / Mr.
Hyde), the twentieth century emphasized utopian social engineering. New
departures in medical research chimed with the new age of technological
progress and provided a context of boundless opportunity for those working within it.[6]

The concept of *usefulness* is the point of contact between human experimentation, knowledge, and the state. It is necessary, therefore, to situate
the relationship between medical science and the individual in the context of a twentieth-century modernity that privileged the body above all
else. Our argument is that in the late modern period, the modern state increasingly used its prerogative to lay claim to the individual body for its
own needs, whether social, economic, or military.[7] Such a claim on the part

of the state and its agents obviously raises the question of consent or contract between it and the individual. And incidentally, it was this relationship that stood at the heart of the deliberations of the prosecution council and judges at Nuremberg and that has been the operative paradigm ever since. In order to historicize human experimentation, we have to move beyond the debate over the issue of "informed consent" as institutionalized through the Nuremberg Code.

In their desire to uncover and explore cases of human experimentation from the latter part of the nineteenth century, historians have begun to piece together a more complicated picture than has hitherto been suspected. Based on individual instances of medical practices involving human experimentation, recent studies have revealed interesting and unsuspected patterns and have raised some new and intriguing historical questions. Medical experimentation in Nazi Germany needs to be flagged here, not so much because of its horrible uniqueness (though not so unique if we consider the Japanese case too), but, as we would argue, because it renders visible what medical historians now know only too well—that such activity has been common to many advanced societies in the twentieth century.[8]

The role of medicine in the Third Reich—the apparent willingness of its practitioners to become accomplices in a crime against humanity—has stood as a warning beacon to the "civilized" world since 1946. Until recently, and largely on account of the Nuremberg trials, historians have tended to approach the history of human experimentation in a number of different ways. One of these has been the uncovering of cases of nontherapeutic experimentation in settings that are wholly unlike those of Nazi Germany. Discovering and exploring instances of such experimentation on human subjects is an important historical exercise. The chapters in this book contribute to this approach. They show clearly how abusive practices could and did flourish in medical practice in situations where the state was not as coercive as in Germany. They also show, in agreement with Susan Lederer, that the Nuremberg Code stands, not as the beginning of a history of human experimentation, but as part of it, and that whether or not such experiments were exceptional and the ethics of them not widely discussed before the 1940s, "understanding the development of ethical standards in the years before World War II is not only important but essential."[9]

But while this volume has consciously avoided a focus on the use of humans in Nazi medical experiments, some discussion of the German context is needed, if only to expose the general problem. We argue that human experimentation as defined at Nuremberg was—and still is—a practice that is not restricted to a particular period or place existing on the other side of the civilizing process. The contributions from Brian Balmer, Glenn Mitchell, and Gilbert Whittemore and Miriam Boleyn-Fitzgerald to this volume expose such activity in the "civilized" democratic societies of Britain, Australia, and the United States.[10] Indeed, it is embedded in the modern tradition.

Another approach to the history of human experimentation, and one that is fully commensurate with what one may call the disclosure perspective, is to place the history of human experimentation at the behest of another project, namely, the history of informed consent. Much recent work and many of the chapters in this book have contributed significantly in this area.[11] However, while we see informed consent as an important area of historical inquiry, we do not agree that "the history of human experimentation cannot be understood independently of the development of ideas about informed consent for medical treatment."[12] One reason is that the historical trajectories and contexts of human experimentation and informed consent are not the same. Human experimentation, as we have stated, is part of the history of medical practice, in particular the scientization and laboratory revolutions of the latter part of the nineteenth century. Another is that focusing on informed consent skews the study of human experimentation toward an ethical analysis rather than a practice. Medical practitioners have adopted informed consent as a means of processing the ambiguity of all research on human subjects. Informed consent does not eliminate that ambiguity, and, we would argue, it obscures rather than helps reveal historical practices. Informed consent is a historical product rather than a tool of historical analysis. Human experiments, even when informed consent has been obtained, may still violate the patient's autonomy.[13] Furthermore, there is the complicating factor that informed consent varies in its meaning from one culture to another, and is therefore not as monolithic as some commentators imply or assume.[14]

The callous treatment of the sick and the horrific experimentation carried out in the medical institutes of the Third Reich's universities, in its hospitals, on the sick wards, in prisons, and, finally, in the concentration camps, led to a setting down of principles intended to regulate the relationship between doctor and patient, science and subject. The key to this new relationship was the recognition of the right of the patient or subject through the principle of "informed consent."[15] This paradigm has dominated discussion since Nuremberg. The overwhelming concentration on the issue of informed consent, with its focus on the relationship between doctors and patients, has, in our view, obscured the important question of the relationship among medical researchers, doctors, and the state as well as between state and society.

Some historians, for example, Detlev Peukert, have recently argued that the progressive secularization of European society from the mid-nineteenth century on was matched by a rising faith in the power of rational science, leading to the emergence of a condition of *logodicy* in which the authority of the church was transferred to science.[16] In early-twentieth-century Germany, for instance, it was widely believed that science could be deployed to resolve social and racial questions.[17] Doctors and biohygienists became the determinators of a bioracially constituted state; they saw themselves as its gatekeepers and guardians, programmed with the mission to secure a utopian healthy society.[18] According to this reading, the unethical experimentation on humans by modern science under the Third Reich represented less its subversion and the compromise of its practitioners than a rational culmination of modernity.[19]

By taking a longer view, we have to ask a different set of questions because, as we know, the scientific impetus to experiment using humans was already there before Hitler and continued after him.[20] For by the beginning of the twentieth century, the boundary between science and the state was becoming progressively blurred as medical men and scientists were absorbed into the wider machinery of the state in ever-increasing numbers. In this process, medical science became a constitutive force in the creation of a "knowledge society" built around the functionality of the body.[21] The substance of the relationship was a combination of scientific technocratism and biologism that focused on the individual as part of the collective or national body.[22] This techno-biological determinism in Germany, as Barondess has argued, was itself part of a wider process that had been clearly evident in Europe and in North America since Darwin.[23] In Germany the process was in full swing by the turn of the century, given a greater impetus by the First World War, consolidated in the decade immediately before 1933, and finally found its apotheosis in the Nazi state.

As historians of medicine have already observed, racial hygiene conflating disease and race was invented by medical science.[24] Medical science thus provided the modern state with a new language that scripted its relationship to society.[25] In this process it was not simply a matter of science serving the state in the utilitarian sense of "what was useful was good," but also of the state serving science.[26] Scientific organizations received their funding from state-sponsored bodies, and their members were often employed in some capacity or other by the state, whether as local medical officers, associates on insurance panels, or members of parliamentary committees of inquiry into public health. In early-twentieth-century Europe, this developing "culture of biologism"[27] was most advanced in Germany, where there were already twenty university institutes and fifteen journals dedicated to the subject.

The most important were the Kaiser Wilhelm Institute for Anthropology in Berlin under Eugen Fischer, the Kaiser Wilhelm Institute for Hereditary Genealogy in Munich under Ernst Rüdin, and (the largest) the Institute for Racial Hygiene in Frankfurt, led by the renowned geneticist Baron Otmar von Verschuer. These institutes, all established before 1933, stood in the vanguard of the state, both under the democratic Weimar Republic and under Nazism. They trained the doctors and medical researchers who eventually found their way to the concentration camps, among them Josef Mengele, who had studied in Munich before becoming von Verschuer's assistant.[28] The prominence given to medical science after 1933 was encapsulated in a description by a leading Nazi, Hans Schemm, of the Nazi state as "applied biology."[29] Indeed, under the Third Reich medical sciences prospered as the state became the chief contracting agent, not only providing finance but also supplying material resources that ultimately were to include human bodies.[30]

These institutes took on a prominence with the outbreak of war in 1939. Through their laboratory work in the death camps, they were able to literally position themselves in the frontline of Germany's racial war in the East. The result was stiffened competition for resources from the key funding body, the German Research Council (Deutsche Forschungsgemeinschaft, or DFG) and a drastic reduction in the number of those now receiving funds. By early 1945, the biosciences located in the Kaiser Wilhelm Institute and the Kaiser Wilhelm Society received somewhere in the region of 80 percent of DFG funding. Scientists had to justify their work in terms of the overall needs of the war effort, and clearly many were well placed to do so, with research ranging from crop experimentation to radiation experiments involving humans.[31] The result was a close affiliation between medical researchers and the state that appears unparalleled in the history of human experimentation. It also produced intense rivalry between institutes and individuals, especially since the careers of many young researchers were closely bound up with the availability of financial resources and the kudos and reputation associated with research that was useful to the state.

Thus, experiments conducted under the auspices of the Nazi state can be better understood when contextualized, without relativizing their horror. Until very recently, historians of Nazi human experimentation have written what we might term an accusatory history: pointing a finger at the crimes of the German medical profession as either a product of something inherently flawed in the German character or as the plaything of an evil racist and totalitarian regime that had collectively abandoned its "civilized" qualities.[32] The vehemence of this approach is understandable, but it diverts us from historicizing human experimentation. When viewed in the specific context of medical and scientific practices of the early twentieth century, medical science under the Third Reich was not sui generis but shared more universal traits than has been hitherto given credence.[33]

The recent research on human experimentation before Nuremberg chimes with Susan Lederer's arguments. In reading through the literature one finds not monsters but medical practitioners who are very human and who are not unlike their colleagues working on different kinds of experiments. As Margaret Humphreys shows in her contribution to this book, Mark Boyd, who made a series of experiments in malariatherapy in the 1930s in Florida for the treatment of syphilis, using inmates of the state mental hospital, struggled with the double-sided aspects of his research: between, on the one hand, his role as a caring physician and, on the other hand, his role as a researcher—a searcher for knowledge with an "experimental gold mine" on his hands. Demonizing Boyd simply pushes away from the central issues. Once we get away from thinking about whether individual practitioners are monsters, we can begin, as does Humphreys, to ask the kind of penetrating questions that make human experimentation such a rich area of investigation. How did Boyd turn his patients into objects? What strategies did he employ within the context of his own career and political milieu? What were the connections between his research results and his research materials, and what reception did such results have? Humphreys provides insightful answers that take us beyond the world of Mark Boyd in the Florida of the 1930s into the broader (and richer) questions of medical practices and human experimentation: Who experimented on whom, on which grounds and conditions, and with which results? Jenny

Stanton's chapter on the work of the Medical Research Council Jaundice Committee in the U.K., and particularly on the work of Fred MacCallum, raises similar issues of retrospective judgment.

In a similar vein, though from a different perspective, the chapter by David Jones and Robert Martensen focuses on human radiation experiments at the University of California both before and after the Second World War. It shows quite vividly the tensions between the wish to exploit new technology and help establish a new medical discipline on the one hand, and ethical issues of experimentation in both nontherapeutic and therapeutic situations on the other. The unethical behavior of the Lawrence brothers, Ernest and John, needs to be understood, Jones and Martensen argue, not as simple transgressions of ethical practice, but rather as a complicated outcome of the power and interest relationships in the laboratory and in the wider scientific world. These are what shaped the research procedures rather than any explicit relationship to the patients. Joel Howell and Rodney Hayward, in this volume, take the analysis a step further by alerting us not only to how experiments were done but to how their stories have been told and re-told.

It is clear that there was a wide disparity between statements of practice and the practices themselves, and between the situation in one country and another. Claude Bernard has often been singled out as constructing an early code for the methods and ethics of experimentation. His statement, written in 1865, that "the principle of medical and surgical morality consists in never performing on man an experiment which might be harmful to him to any extent, even though the result might be highly advantageous to science, i.e., to the health of others," has often been quoted as an example of an early coherent statement on ethics, but it was not a reflection of what actually happened.[34] Bernard was not alone in stating or discussing the ethics of experimentation, as the examples of Thomas Percival in England and William Beaumont in the United States make clear.[35] But practices were different, and the impetus to provide codes of conduct in the clinic and the laboratory did not, on the whole, come from the practitioners themselves.

Germany, or more specifically, Prussia, was the first state to regulate nontherapeutic research. The initiative did not come from the medical profession but from the state. The earliest directive, issued by the Prussian minister of the interior in 1891, was quite narrow in its scope, with a view to regulating therapeutic practices in Prussian prisons—specifically, advising practitioners that they could not administer tuberculin for the treatment of tuberculosis if it was against the patient's will.[36] This was followed in 1900 by a more detailed set of regulations issued by the Prussian minister for religious, educational, and medical affairs, who stipulated that nontherapeutic experiments in hospitals and clinics could only be undertaken by the medical director and that the human subjects had to give unambiguous consent after being informed of the possible risks of sustaining damage or even death as a result of the experiment.[37] This ministerial directive came directly in response to the controversy surrounding Dr. Albert Neisser's well-publicized 1898 experiments on syphilis prevention undertaken on women patients (many of whom were prostitutes), none of whom had been asked for their consent or been informed of the risks involved, and who contracted syphilis because of the experiment. While it was enlightened, the directive was not, however, legally binding.[38]

Concern over using humans in medical trials continued into the 1920s and was kept in the public domain through ongoing debate in the Reich parliament. While it was clear that the medical profession could not regulate itself adequately, there was little immediate progress toward bringing its practices under some form of state control. This only came in a rather dramatic way at the beginning of the 1930s after a particularly shocking case came to light that recalled the Neisser clinical trials of the 1890s. The event that triggered public outcry and spurred the government into action occurred at a clinic in Lubeck, where almost one-third of the children used to test a new vaccine against tuberculosis died.[39] The resultant legislation in 1931 issued detailed guidelines for therapeutic and nontherapeutic procedures, thereby clearly distinguishing for the first time between treatment and experiment. The consensus of historians and ethicists seems to be that this code was the most comprehensive ever seen and that in many ways it was stricter than the subsequent codes of Nuremberg and Helsinki.[40] After 1933 this constraint upon medicine and its relationship to the individual and the state was conveniently ignored, though it remained on the statute books.[41]

As far as recent research has been able to tell, advising or regulating medical practices concerned with experimentation elsewhere in Europe and in the United States was mostly lacking. In France, the issue of human experimentation was dominated by the work and personality of Louis Pasteur, whose research into an anti-rabies vaccine and the ethics surrounding it aroused interest in France and elsewhere. Pasteur's practices were criticized by medical practitioners at the time, who pointed out that he had exaggerated the amount of animal experimentation undertaken before treating Joseph Meister, the nine-year-old from Alsace, his first subject.[42] Criticism notwithstanding, the voices ranged against Pasteur (and others) did not manifest themselves in any attempt to regulate practices and were soon forgotten. What survived was the example of Pasteur, in particular his drive and exuberance, reflected in his methods and embraced by devotees such as Waldemar Haffkine, whose work on an anti-cholera vaccine toward the end of the nineteenth century in France and India also appears to have taken liberties in the pre-human experimental stages.[43]

In the United States, the situation was again different. Nontherapeutic experiments providing medical knowledge rather than directly benefiting patients emerged, as they did in both Germany and France, as part of medical practice toward the end of the nineteenth century when medicine itself began to embrace experimental science. The most wide-ranging discussion of these practices in the United States appears in a very important book by Susan Lederer, *Subjected to Science* (1995).[44] Lederer is not concerned specifically with the ethics of the practices. Rather, she is concerned with human experimentation as a vehicle for discussing medical science and its transformation from the late nineteenth century to World War II. What she discovers is somewhat surprising. Experimental encounters, as she refers to them, were complex phenomena. Medical practitioners discussed the ethics of their practices openly, she argues, but as elsewhere, talk did not lead to regulating experimentation with humans. To understand why regulation did not follow, Lederer introduces us to an unlikely group of protesters, those who were against animal vivisection, and shows convincingly how their politics shaped the politics of medical research practices.

Lederer bases her conclusions on a number of case studies, notably experiments at the University of Pennsylvania, the University of Michigan, and the Rockefeller Institute, and the Tuskegee trials in the first few decades of the twentieth century.[45] The "investigative materials" included orphans, soldiers, prisoners, patients in mental hospitals and African Americans. These experiments attracted the attention of antivivisection activists whose field of action had now broadened to include humans as well as animals.[46] Particularly effective were a number of prominent women who attacked male medical researchers as heartless and more like children tormenting animals than caring adults. The practitioners, for their part, reacted with arguments as well as with practice, resorting to self-experimentation in order to underline the heroism and martyrdom of modern medical scientists. The important point here is that Lederer would have missed this connection entirely had she focused on the narrow issue of whether or not such and such experiment was ethical. Instead, we are offered a political history of humans as "investigative material"; that is, a discourse on humans simultaneously as objects and subjects within a transforming medical science.

The vexed question of the past role of doctors and medical researchers in human experimentation turns on the idea that there has been a betrayal of the Hippocratic Oath.[47] The assumption is that medical science is there to provide healing remedies for the sick individual, when in fact its role in the modern era—indeed, ever since Jenner's experiments with smallpox vaccination in 1798—has been to safeguard the collective national health. More specifically, from the beginning of the twentieth century racial hygienists, medical doctors, and scientists have sought to reconstitute the individual body as a healthy part of a "resilient" national body.[48] In this paradigm, the post-Hippocratic body is stripped of its individuality and subsumed into the larger "personality" of the national body that is *racially* constituted.[49]

In Germany after the First World War, doctors faced the twofold task of healing the sick (and defeated) nation after the ravages of war and revolution, and of regenerating it for future fitness, transforming it into a productive force capable of defending itself in a hostile and uncertain world. They were, accordingly, not only the gatekeepers of the nation's health but also its physiological engineers. We might go further and say that the "nation" was their laboratory and that they became the new gods of science, creators of life that was useful.[50] In part this explains their apparent inability to relate to their subjects where issues of racial hygiene and national well-being were concerned.[51] How could they empathize with subjects they perceived as uncivilized, undisciplined, unhealthy, dysfunctional, irrational, unproductive, primitive, degenerate, and impulsive?[52] Instead, doctors identified with the state they served and were shaping. Their primary aim was to heal the national body so that it could endure eternally. But the route to this goal could only be achieved by transforming the individual body, itself ephemeral and interchangeable—and ultimately disposable.[53] This undertaking involved a process of experimentation. Hospitals, mental institutions, and (unpalatable as it may appear) the Nazi concentration camps, became sites for the repression of biological crisis, while at the same time functioning as laboratories for the production of the racially perfected national body.

Through medical experimentation, use*less* bodies were rendered use*ful* by being made us*able* in the national project of regeneration, thus gaining a utility they were believed otherwise to lack. An example of this can be found in the so-called Schaltenbrand Experiments of 1940, where the senior doctor at Wurzburg University's medical clinic, the distinguished neuroscientist Georg Schaltenbrand, had been experimenting on monkeys in his determination to establish a viral etiology for multiple sclerosis (MS). In order to progress further with his research, Schaltenbrand decided that he needed to move beyond monkeys to see whether he could induce MS in humans under clinical conditions. In spite of an apparently low risk of actually succeeding, Schaltenbrand argued the necessity for carrying out experiments on patients deemed "useless imbeciles," rather than "wasting" healthy and thus "valuable" human volunteers.[54] In all, he experimented on forty-five patients, including children, of whom at least two died and others suffered a range of physical and mental disorders as a result. These were the only conclusive results from the experiment. Not surprisingly, the viral etiology continued to elude him.[55]

Using bodies in such a way made sense to these doctors. As they had in peacetime, they now in a time of national emergency placed their expertise at the disposal of the state. The Hippocratic Oath was simply not an issue here, when the "life of the nation" was at stake. This removed any moral dilemma that might be associated with experimenting on humans. Either as "gods of science" they were beyond morality, or as mere laboratory workers they felt that responsibility lay further up the line in the vague and ill-defined interstices of the bureaucratic "agentic state."[56]

Not all experiments conducted by doctors such as that carried out by Schaltenbrand were fraudulent practices. In trying to explain Nazi medicine's utter disregard for human life, Leo Alexander, the expert medical witness for the prosecution at the Doctors' Trial in 1946, argued that a "Hegelian utility had displaced moral, ethical religious values" in Nazi Germany.[57] But Alexander himself could not escape the very same utilitarianism—indeed, it underwrites the Nuremberg Code itself! At the end of his report on the human threshold experiments carried out by doctors in Dachau and Auschwitz for the Luftwaffe, Alexander commented that such experiments had their uses, even those involving death and individual suffering, such as Sigmund Rascher's cold water and altitude experiments on prisoners at both concentration camps, and might be permissible in the context of a national emergency (if based on consent!).[58] Alexander clearly did not see the irony in his words, and indeed, he went on to build his own career on the basis of such findings.[59]

The key concept in Alexander's approach to the question of Nazi experiments was that of *consent*. And it is this concept that has governed the debates ever since. What this focus does, of course, is to divert attention away from the issue of experiments per se. It allows for humans to be used in the name of science in its quest to "discover the secrets of nature," provided that they are willing scientific objects.[60] But the individual or collective decision to participate in an experiment might be founded on misinformation, misplaced trust in the profession, or pecuniary circumstances. Meanwhile, the consent of military personnel—and, with clear echoes of Nazi Germany, of prisoners—may not even be considered necessary, as recent declassified Pentagon papers from the 1950s have revealed.[61]

However, the role of the state in experimentation became less clearly visible in the second half of the twentieth century. Increasingly, research became "de-centered" as it became more commercialized, and moved beyond the immediate sphere of the state or state-related agencies and transcended national borders, borne on the wings of multinational corporations. For opponents to using humans in experimental research, finding a target for aiming ethical questions was no longer as simple as Nuremberg had been. Moreover, and apparently unshaken by the implications of Nazi medical research, postwar experimentation was able to reconnect to the positivist traditions of science and modernity with apparent ease.[62]

Thus, while some efforts to regulate human experimentation and enforce ethical medical practice increased in the years following the Second World War, this did little to halt the growth of funding for human experimentation. In one American hospital, Massachusetts General, for example, the amount of money for research grew nearly twenty-fold, from $500,000 in 1945 to more than $8 million in 1965. Central funds for research also rose greatly during this same period, with the grants from the National Institutes of Health (NIH) increasing more than sixty times to a figure exceeding $430 million.[63]

Such increases in state funding of medical research coincided with a new sense of hope and promise in the wake of the development of antibiotics. This was especially noticeable in the United States and in Britain, unencumbered as they were by the taint of negative medicine as in Germany and in Japan. Sparking a belief that medical science could ultimately defeat disease, the remarkable success of antibiotics spurred on a search for new drugs and therapies worldwide, resulting in a large expansion of the pharmaceutical industry. In the 1950s and 1960s, investment in new medical innovations soared, and sales of pharmaceutical drugs grew rapidly. By the mid-1950s the American pharmaceutical industry was marketing more than four hundred new drugs every year, and the number of prescriptions had almost quadrupled since the 1930s. Similarly, drug exports from the United States increased nearly twenty-fold between the end of the Second World War and the 1950s.[64]

In the drive to advance medical knowledge and treatment, the demand for human experiments grew, and the types of experiments that were performed changed. This not only posed new ethical questions about the use of humans for experiments, but it also raised important questions about evaluating the safety and efficacy of the new pharmaceutical drugs and treatments coming on to the market. By the late 1950s many medical practitioners and government officials in Britain and the United States were beginning to express doubts about the quality of regulations then in place for monitoring safety. As in the case of interwar concerns over experimental medicine, it took the human tragedy surrounding the thalidomide scandal (1959–61) to spur governments into action.

Developed by a newly established East German company, Chemie Grünenthal, thalidomide had been promoted as a drug that was "as safe as a mother's milk" and had become widely prescribed in Europe as a sedative and treatment for morning sickness among pregnant women. Thalidomide had, in fact, been considered so safe a drug that it could be obtained without a prescription in Germany. Soon after its appearance, however, the drug was quickly withdrawn from the European market when it was shown to have caused an epidemic of severe birth defects, including phocomelia (lack of fully developed limbs). By 1961 at least 10,000 children worldwide had been born with deformed limbs as a result of the drug, and at least 4,000 had died as a result. Thalidomide came to epitomize not only the potential and unknown dangers posed by any drug used in pregnancy but also the hazards of drugs overall.[65]

Following the thalidomide disaster and amidst a great pressure for change, lengthy criticisms were leveled at regulators for protecting manufacturers and maintaining secrecy policies to the point of absurdity. Such criticism led to hearings in the United States that revealed the process by which several new drugs, originally approved around this period, were removed from the market because they posed what everyone agreed were unacceptable dangers to public health. The horror that thalidomide inspired led directly to stronger laws governing the marketing of new drugs in most of Europe between 1962 and 1964. This brought to an end the quasi-laissez-faire approach that had existed until then.[66] Stricter rules governing the introduction of new drugs were also passed in the United States, where a license for the drug narrowly missed being granted.[67] These rules augmented the strict guidelines that had been introduced almost two decades earlier in the wake of the sulfa drug experiment that had resulted in 109 deaths in 1937.[68]

Paradoxically, the new regulations introduced in the United States and elsewhere in the 1960s increased the amount of testing new drugs now had to undergo in both animals and humans before they could be approved for market. The new legislation also demanded new standards in the conduct of clinical trials. In the United States, for example, the regulations made it explicit for the first time that patients had to be told of the experimental nature of any new drug they were given. Stricter measures were introduced for testing drugs in their preclinical and clinical phase, and regulatory authorities increasingly encouraged larger trials in the belief that this provided greater statistical confidence in the results.[69]

By the end of the twentieth century a shift in the geographical location of bodies considered useful for nontherapeutic research had occurred, indicating not only the globalization of the pharmaceutical market but also that the pharmaceutical industry can actively evade controls over their activities in those states with a developed public sphere. Indeed, in some cases, states may even encourage such a move by facilitatory measures. Thus, part of the growth in the number of subjects and trials in countries outside the United States can be accounted for by the fact that from 1987 the Food and Drug Administration (FDA) allowed new drug applications to be based solely on foreign data.[70] Of those enrolled for trials with antimicrobial drugs in the years between 1978 and 1990, for example, the percentage of patients tested in foreign countries increased from 34 to 41 percent. Meanwhile, over the same period the percentage of American patients enrolled dropped from 66 to 51 percent.[71]

The sheer range of experiments being conducted makes it difficult to estimate the precise number of humans currently being experimented on. Since the Second World War, however, human experimentation has increased enormously in quantity, scale, and type, including not only using whole human bodies but also parts of bodies. Today research conducted ranges from testing new technologies and therapeutic and pharmaceutical interventions, to studying the mechanisms of human disease, to epidemiological and behavioral research, to outcome studies and the evaluation of health services. Much of this research involves direct interaction with the subjects, but experiments are also being done on material of human origin such as cells, tissues, and other specimens, where there is no direct involvement of the subject. Indeed, the establishment of the human genome project has increased this type of research in recent years.

Today, unlike during the earlier periods in the history of human experimentations, the concept of "useful," like the boundary between private and public and between corporation and state, has become blurred. Whereas for most of the twentieth century medical science focused its efforts on experimenting on "useless" bodies, hidden from public view in closed institutions, today the site of research has moved from the confines of the hospital and prison to random populations at large. And as we can see from the foregoing discussion of the pharmaceutical industry, these populations need not be assembled in any particular site but can be scattered across the world.

A further difference between now and then is that experimental subjects can no longer be viewed simply as passive patients selected by doctors, but they themselves have begun to demand entrance to clinical trials. Thus, defining the "useless" and the "useful" became a mutual process between experimenter and subject. Part of this stems from the activism of AIDS patients in the late 1980s and 1990s for wider access to clinical trials in order to try out drugs not yet available on the market.[72] This changed the whole nature of recruitment for clinical trials. This situation was reinforced in the late 1990s as the prerogative of the state to "gate-keep" clinical trials was challenged by the appearance of Web sites advertising clinical trials for various diseases in the United States and elsewhere whereby patients suffering from a diversity of diseases can potentially volunteer for trials promoted on the Internet. To some extent the expansion of self-enrollment is new. If this trend has any resonance with the past, it might be seen in the recruitment of volunteers in prisons and military camps. The motivations of prisoners and military recruits, however, differed greatly from those of people volunteering to take part in the clinical trials of today.

As we have been arguing, the question of "informed consent" does not really arise here. What has actually been at issue throughout the recent history of using humans in the service of medicine is the specific context and the mechanisms that condition the parameters of the experiment. For much of the twentieth century, these have been determined either actively or indirectly by the state. Even at the end of the twentieth century, in a condition of "post-state," it is the state that still largely procures the bodies, because ultimately, bodies are the only possession still within its control. Indeed, a recent road speed experiment in Israel, with deadly consequences, involved the country's entire driving population.[73] Consent was not sought because the Ministry of Transportation, like many state agencies, probably viewed the bodies to be used as necessary, and thus "useful." Nor were controls put in place—and, inevitably, deaths resulted. It is a particular irony that this experiment, though not strictly speaking a medical one, bore many of the characteristics of the attitude shown toward human life in 1930s Germany.

In his 1924 novel *The Magic Mountain*, Nobel Prize winner Thomas Mann was sharply critical of the notion of the pre-Hippocratic body that was defined simply in terms of its usability or nonusability.[74] Yet what he saw as antihuman was in tune with eugenic thought at the time and has been broadly accepted by the medical and scientific community since then.[75] The emergence of the socio-biological trope of the healthy body has thus signaled a fundamental shift in social power.

As the contributions to this volume amply demonstrate, the body has not only become the object of scientific enquiry, but it has also been seen as a raw material to be configured into its final and useful form. This final shape of human usefulness did not have universal characteristics, but was determined by a specific historical context, itself dependent upon contingent factors. Viewed in this way, using bodies can be both historicized and understood. Contemporary practices can only benefit from the kind of historical understanding we are presenting, an understanding that seeks neither to demonize the past nor to celebrate the present. None of the issues raised in this and in subsequent chapters have disappeared.

NOTES

1. William Bynum, "Reflections on the History of Human Experimentation," in *The Use of Human Beings in Research*, ed. Stuart F. Spicker, Ilai Alon, Andre de Vries, and H. Tristram Engelhardt Jr. (Dordrecht, 1988), 30.

2. Bynum (n. 1 above), 31. Bynum's typology is based on five idealizations: bedside, library, hospital, laboratory, and social medicine, covering two and a half millennia of medical history.

3. Gert H. Brieger, "Human Experimentation: History," in *Encyclopedia of Bioethics*, ed. Warren Thomas Reich (New York, 1978), 684.

4. Both animal experimentation and self-experimentation could do with substantially more research both in their own right and in underpinning research into human experimentation. On animal experimentation or vivisection, see Nicholas A. Rupke, ed., *Vivisection in Historical Perspective* (London, 1987); Andreas-Holger Maehle, "The Ethical Discourse on Animal Experimentation, 1650–1900," in *Doctors and Ethics: The Earlier Historical Setting of Professional Ethics*, ed. Andrew Wear, Johanna Geyer-Kordesch, and Roger French, Clio Medica 24 (Amsterdam, 1993), 203–51; and R. Singleton Jr., "Whither Goest Vivisection? Historical and Philosophical Perspectives," *Perspectives in Biology and Medicine* 37 (1993–4): 576–94. On self-experimentation, a good place to start is with Lawrence K. Altman, *Who Goes First? The Story of Self-Experimentation in Medicine* (Berkeley, CA, 1998).

5. David J. Rothman, "Research, Human: Historical Aspects," in *Encyclopedia of Bioethics*, ed. Warren Thomas Reich (New York, 1995), 2249.

6. Christian Pross, "Nazi Doctors, German Medicine, and Historical Truth," in *The Nazi Doctors and the Nuremberg Code: Human Rights in Human Experimentation*, ed. George J. Annas and Michael A. Grodin (Oxford and New York, 1992), 39; Michael A. Grodin, "Historical Origins of the Nuremberg Code," in ibid., 124.

7. Michael Marrus, "The Nuremberg Doctors' Trial in Historical Context," *Bulletin for the History of Medicine* 73 (1999): 122.

8. See Paul Weindling, "Human Guinea Pigs and the Ethics of Experimentation: The *BMJ*'s Correspondent at the Nuremberg Medical Trial," *British Medical Journal* 313 (7 Dec. 1996): 1467–70. For Japan, see Hal Gold, *Unit 731 Testimony* (Tokyo, 1996); Sheldon H. Harris, *Factories of Death: Japanese Biological Warfare, 1932–45, and the American Cover-up* (London, 1994); and Sheldon H. Harris, "The American Cover-up of Japanese Human Biological Warfare Experiments, 1945–1948," in *Science and the Pacific War: Science and Survival in the Pacific, 1939–1945*, ed. Roy M. MacLeod (Dordrecht, 2000), 253–69.

9. Susan E. Lederer, *Subjected to Science: Human Experimentation in America Before the Second World War* (Baltimore, MD, 1995), xiii. The point about the paucity of experimentation on humans before World War II was made in David J. Rothman, "Ethics and Human Experimentation," *New England Journal of Medicine* 317 (5 Nov. 1987): 1196.

10. For Australia, see Bridget Goodwin, "Australia's Mustard Gas Guinea Pigs," in *Science and the Pacific War: Science and Survival in the Pacific, 1939–1945*, ed. Roy M. MacLeod (Dordrecht, 2000), 139–71.

11. See, for example, Weindling (n. 8 above), and Ruth R. Faden and Tom L. Beauchamp, *A History and Theory of Informed Consent* (New York, 1986).

12. Lederer (n. 9 above), 13.

13. See, for example, the excellent discussion in Dean Cocking and Justin Oakley, "Medical Experimentation, Informed Consent and Using People," *Bioethics* 8 (1994): 293–311. Cocking and Oakley are particularly concerned with clinical trials of new therapies, a point discussed later in this introduction.

14. George J. Annas and Frances H. Miller, "The Empire of Death: How Culture and Economics Affect Informed Consent in the U.S., the U.K., and Japan," *American Journal of Law & Medicine* 20 (1994): 357–94.

15. See the special issues of *British Medical Journal* 1996 and *JAMA* 1996; George J. Annas and Michael A. Grodin, eds., *The Nazi Doctors and the Nuremberg Code: Human Rights in Human Experimentation* (New York, 1992); and John J. Michalczyk, ed., *Medicine, Ethics, and the Third Reich: Historical and Contemporary Issues* (Kansas City, MO, 1994).

16. Detlev Peukert, "The Genesis of the 'Final Solution' from the Spirit of Science," in *Reevaluating the Third Reich*, ed. Tom Childers and Jane Caplan (New York, 1993), 241.

17. Anthony McElligott, *The German Urban Experience 1900–1945: Modernity and Crisis* (London, 2001), chap. 5.

18. Allan Fenigstein, in *Lessons and Legacies: Teaching the Holocaust in a Changing World II*, ed. Donald G. Schilling (Evanston, IL, 1998), 70; Robert N. Proctor, "Nazi Doctors, Racial Medicine and Human Experimentation," in Annas and Grodin (n. 15 above), 19.

19. Zigmunt Bauman, *Modernity and the Holocaust* (Cambridge, 1989). In a similar vein, see Jochen Vollmann and Rolf Winau, "Informed Consent in Human Experimentation before the Nuremberg Code," *British Medical Journal* 313 (7 December 1996): 1445–7; and Mark Walker, "National Socialism and German Physics," in *Journal of Contemporary History* 24 (1989): 85.

20. William E. Seidelman, "Mengele Medicus: Medicine's Nazi Heritage," *Millbank Quarterly* 66 (1988): 221–239.

21. McElligott (n. 17 above) and Margit Szöllösi-Janze, "National Socialism and the Sciences: Reflections, Conclusions and Historical Perspectives," in *Science in the Third Reich*, ed. Margit Szöllösi-Janze (Oxford, New York, 2001), 2ff. 8–10, 17–19.

22. G. von Schulze-Gävernitz, "Die Maschine in der kapitalistischen Wirtschaftsordnung," *Archiv für Sozialwissenschaft und Sozialpolitik* 63 (1930): 225–73; Nikolas Rose, "Medicine, History and the Present," in *Reassessing Foucault: Power, Medicine and the Body*, ed. Colin Jones and Roy Porter (London, 1994), 70.

23. Jeremiah A. Barondess, "Medicine Against Society: Lessons from the Third Reich," *Journal of the American Medical Association* 276 (27 Nov. 1996): 1660. For Germany, see Volker Roelcke, "Using Bodies in a Culture of Biologism: Psychiatric Research in Germany, 1933–1945," conference paper presented at *Using Bodies: Humans in the Service of 20th Century Medicine*, London, 3–4 September 1998.

24. Robert J. Lifton, *The Nazi Doctors: Medical Killing and the Psychology of Genocide* (New York, 1986), 3.

25. Pross (n. 6 above), 35.

26. Josef Reindl, "Believers in an Age of Heresy? Oskar Vogt, Nikolai Timofé-eff-Ressovsky and Julius Hallervorden at the Kaiser Wilhelm Institute for Brain Research," in Szöllösi-Janze (n. 21 above), 234.

27. On this phenomenon see Volker Roelcke, "Biologizing Social Facts: An Early 20th Century Debate on Kraepilin's Concept of Culture, Neurasthenia, and Degeneration," *Culture, Medicine and Psychiatry* 21 (1997): 383–403; Roelcke (n. 23 above).

28. Proctor (n. 18 above), 20.

29. Max Weinreich, *Hitler's Professors: The Part of Scholarship in Germany's Crimes Against the Jewish People* (orig. 1946, new edition with an introduction by Martin Gilbert) (New Haven, CT, 1999), 34; Peukert (n. 16 above), 235ff.

30. Ute Deichmann and Benno Müller-Hill, "Biological Research at Universities and Kaiser Wilhelm Institutes in Nazi Germany," in *Science, Technology, and National Socialism*, ed. Monika Renneberg and Mark Walker (Cambridge, 1994), 160–83.

31. Deichmann and Müller-Hill (n. 30 above), 167–75; Ute Deichmann, *Biologists Under Hitler* (Cambridge, MA, 1996)

32. As representative of this approach, see: Götz Aly, Peter Chroust, and Christian Pross, *Cleansing the Fatherland: Nazi Medicine and Racial Hygiene* (Baltimore, MD, 1994). See also Seidelman (n. 20 above), 228–29.

33. Pross (n. 6 above), 39.

34. Claude Bernard, *An Introduction to the Study of Experimental Medicine* (New York, 1957 [1865]), 101.

35. Thomas Percival, *Medical Ethics* (Manchester, 1803). For a recent discussion of Percival, see Robert Baker, Dorothy Porter, and Roy Porter, eds., *The Codification of Medical Morality: Historical and Philosophical Studies of the Formalization of Western Medical Morality in the Eighteenth and Nineteenth Centuries* (Dordrecht, 1993); and Lisbeth Haakonssen, *Medicine and Morals in the Enlightenment: John Gregory, Thomas Percival, and Benjamin Rush* (Amsterdam, 1997). On Beaumont, see Ronald L. Numbers, "William Beaumont and the Ethics of Human Experimentation," *Journal of the History of Biology* 12 (1979): 113–35.

36. Vollmann and Winau (n. 19 above), 1445.

37. Ibid., 1446.

38. On the Neisser case, see Barbara Elkeles, "Medizinische Menschenversuche gegen Ende des 19. Jahrhunderts und der Fall Neisser. Rechtfertigung und Kritik einer wissenschaftlichen Methode," *Medizinhistorisches Journal* 20 (1985): 135–48; and Lutz Sauerteig, "Ethische Richtlinien, Patientrechte und Ärztliches Verhalten bei der Arzneimittelerprobung (1892–1931)," *Medizinhistorisches Journal* 35 (2000): 303–34.

39. On the Lubeck episode, see P. Menut, "The Lubeck Catastrophe and Its Consequences for Anti-Tuberculosis BCG Vaccination," in *Singular Selves. Historical Issues and Contemporary Debates in Immunology*, ed. Anne-Marie Moulin and Alberto Cambrosio (Amsterdam, 2001), 202–10; and Christian Bonah, "Ethique et recherche biomédicale en Allemagne: le process de Lubeck et les Ricthlinien de 1931," *Journal Internationale de Bioéthique* 12 (2001): 23–39. On the 1931 code, see

Vollmann and Winau (n. 5 above), 1446–47; Norman Howard-Jones, "Human Experimentation in Historical and Ethical Perspectives," *Social Science and Medicine* 16 (1982): 1435–36; Paul Weindling, "The Origins of Informed Consent: The International Scientific Commission on Medical War Crimes, and the Nuremberg Code," *Bulletin of the History of Medicine* 75 (2001): 41–42; Robert N. Proctor, "Nazi Science and Nazi Medical Ethics: Some Myths and Misconceptions," *Perspectives in Biology and Medicine* 43 (2000): 342.

40. See, for example, Howard Jones (n. 39 above), 1436.

41. Peukert (n. 16 above), 244.

42. Gerald L. Geison, "Pasteur's Work on Rabies: Reexamining the Ethical Issues," *Hastings Center Report* (April 1978): 26–34; and Gerald L. Geison, "Pasteur, Roux, and Rabies: Scientific *versus* Clinical Mentalities," *Journal of the History of Medicine and Allied Sciences* 45 (1990): 341–65.

43. Ilana Löwy, "From Guinea Pigs to Man: The Development of Haffkine's Anticholera Vaccine," *Journal of the History of Medicine and Allied Sciences* 47 (1992): 270–309. For the case of another practitioner, see Maurice Huet, "L'expérimentation humaine au temps de Charles Nicolle," *Histoire des sciences médicales* 34 (2000): 409–14. On the debate surrounding syphilization, see Joan Sherwood, "Syphilization: Human Experimentation in the Search for a Syphilis Vaccine in the Nineteenth Century," *Journal of the History of Medicine and Allied Sciences* 54 (1999): 364–86.

44. Susan E. Lederer, *Subjected to Science: Human Experimentation in America before the Second World War* (Baltimore, MD, 1995).

45. Earlier versions of some of the case studies in Lederer's book can be found in Susan E. Lederer, "The Right and Wrong of Making Experiments on Human Beings: Udo J. Wile and Syphilis," *Bulletin of the History of Medicine* 58 (1984): 380–97; and "Hideyo Noguchi's Luetin Experiment and the Antivivisectionists," *Isis* 76 (1985): 31–48. See also the excellent study on the use of slaves as human subjects in medical experiments in Todd L. Savitt, "The Use of Blacks for Medical Experimentation in the Old South," *Journal of Southern History* 48 (1982): 331–48; and Durrinda Ojanuga, "The Medical Ethics of the Father of Gynecology, Dr. J. Marion Sims," *Journal of Medical Ethics* 19 (1993): 28–31. The classic study of the infamous Tuskegee experiment is James H. Jones, *Bad Blood: The Tuskegee Syphilis Experiment*, new and expanded edition (New York, 1993). The literature on Tuskegee is large and growing. In addition to Jones, see recent contributions giving greater bibliographical coverage: Thomas G. Benedek and Jonathan Erlen, "The Scientific Environment of the Tuskegee Study of Syphilis, 1920–1960," *Perspectives in Biology and Medicine* 43 (1999): 1–30; Susan M. Reverby, "Rethinking the Tuskegee Syphilis Study: Nurse Rivers, Silence and the Meaning of Treatment," *Nursing History Review* 7 (1999): 3–28; and Susan M. Reverby, ed., *Tuskegee's Truths: Rethinking the Tuskegee Syphilis Study* (Chapel Hill, NC, 2000). Another infamous medical experiment is covered in William B. Bean, "Walter Reed and the Ordeal of Human Experiments," *Bulletin of the History of Medicine* 51 (1977): 75–92.

46. There is a great more study to be done on animal vivisection and its relationships, if any, to criticisms of human experimentation. In addition to Lederer's book, see the contributions cited in note 4 above.

47. Michael A. Grodin, George J. Annas, and Leonard H. Glantz, "Medicine and Human Rights. A Proposal for International Action," *Hastings Center Report* 23, no. 4 (1993): 8.

48. Professor Dr. Nowack, "Die öffentliche Gesundheitspflege," in *Die Deutschen Städte: Geschildert nach den Ergebnissen der ersten deutschen Städteausstellung zu Dresden 1903*, ed. Robert Wuttke (Leipzig, 1904), 446.

49. For an interesting discussion, see Paul Weindling, *Health, Race and German Politics Between National Unification and Nazism, 1870–1945* (Cambridge, 1989).

50. A brilliant example of this is given in Fritz Lang's film *Metropolis* (1926), where the scientist Rotwang creates the robot Maria, who will become the ultimate producer and reproducer of goods and life.

51. Lifton (n. 24 above), 611; and Fenigstein (n. 18 above), 62–65.

52. McElligott (n. 17 above), 80–81, 109.

53. Peukert (n. 16 above), 241.

54. The idea of "uselessness" underlies much of the choice of investigative subjects. For much of the twentieth century, the use of the experiment as a means of changing useless to useful revolved around using prisoners, orphans, and hospital inmates as material. On prisoners as human subjects, see Allen M. Hornblum, "They were Cheap and Available: Prisoners as Research Subjects in Twentieth Century America," *British Medical Journal* 315 (29 Nov. 1997): 1437–41; Allen M. Hornblum, *Acres of Skin* (New York, 1998); and Jon M. Harkness, "Research Behind Bars: A History of Nontherapeutic Research on American Prisoners," Ph.D. dissertation, University of Wisconsin, 1996. On orphans, see Susan E. Lederer, "Orphans as Guinea Pigs: American Children and American Experimenters, 1890–1930," in *In the Name of the Child: Health and Welfare, 1880–1940*, ed. Roger Cooter (London, 1992), 96–123; and Susan E. Lederer and Michael A. Grodin, "Historical Overview," in *Children as Research Subjects: Science, Ethics, and Law*, ed. Michael A. Grodin and Leonard H. Glantz (New York, 1994), 3–28. The use of military personnel was a more complicated story, but see Bean (n. 44 above) and Susan E. Lederer, "Military Personnel as Research Subjects," in *Encyclopedia of Bioethics*, ed. Warren Thomas Reich (New York, 1995), 1774–76.

55. M. I. Shevell and B. K. Evans, "The 'Schaltenbrand Experiment' Wurzburg 1940: Scientific, Historical, and Ethical Perspectives," *Neurology* 44 (1994): 351–52.

56. Fenigstein (n. 18 above), 60, 62, 69; and Raul Hilberg, *The Destruction of European Jews* (Chicago, 1961).

57. Barondess (n. 23 above), 1658.

58. International Military Tribunal, *Trial of the Major War Criminals before the IMT: Nuremberg 14 November 1945–1 October 1946*, MT 400–PS Excerpts from Alexander's report and appendices, 400–PS 5.

59. Seidelman (n. 20 above), 229–30.

60. A. C. Ivy, "The History and Ethics of the Use of Human Subjects in Medical Experiments," *Science* 108 (2 July 1948): 1–5.

61. Jonathan D. Moreno, "'The Only Feasible Means': The Pentagon's Ambivalent Relationship with the Nuremberg Code," *Hastings Center Report* 26, no. 5 (1996): 11–19.

62. Grodin, Annas, and Glantz (n. 47 above), 10.

63. Henry K. Beecher, "Ethics and Clinical Research" *New England Journal of Medicine*, 274 (1966): 1355.

64. By the end of the 1950s there were 12,000 pharmaceutical manufacturing companies in the United States and 56,000 retail drug stores. See T. Stephens and R. Brynner, *Dark Remedy: The Impact of Thalidomide and Its Revival as a Vital Medicine* (Cambridge, MA, 2001), 14, 101.

65. H. Sjöström and R. Nilsson, *Thalidomide and the Power of the Drug Companies* (London, 1972); Stephens and Brynner (n. 64 above).

66. Sjöström and Nilsson (n. 65 above) and John Abraham, *Science, Politics, and the Pharmaceutical Industry: Controversy and Bias in Drug Regulation* (London, 1995), 36–86.

67. Linda Bren, "Frances Oldham Kelsey: FDA Medical Reviewer Leaves Her Mark on History," *FDA Consumer*, 35 (2001): 24–29; R. McFadyen, "Thalidomide in America: A Brush with Tragedy," *Clio Medica* 11 (1976): 79–93.

68. C. Jackson, *Food and Drug Legislation in the New Deal* (Princeton, 1970); J. H. Young, "Sulfanilamide and Diethylene Glycol," in *Chemistry and Modern Society*, ed. J. Parascandola and J. Whorton (Washington, DC, 1983), 105–25; Senate Committee on Government Operations, *Hearings on Interagency Coordination in Drug Research and Regulations, Part 3*, March 1963, 987; Thomas Maeder, *Adverse Reactions* (New York, 1994), 125–51.

69. Office of Technology Assessment, *Pharmaceuticals: Costs, Risks and Rewards* (February 1993), OTN-H-522, 146.

70. Ibid., 146.

71. Ibid., 14, 146.

72. Steven Epstein, *Impure Science: AIDS, Activism and the Politics of Knowledge* (Berkeley, CA, 1996); Steven Epstein, "The Construction of Lay Expertise—AIDS Activism in the Reform of Clinical Trials," *Science, Technology & Human Values* 20 (1995): 408–37; and Steven Epstein, "Activism, Drug Regulation, and the Politics of Therapeutic Evaluation in the AIDS Era: A Case Study of ddC and the 'Surrogate Markers' Debate," *Social Studies of Science* 27 (1997): 691–726.

73. Elihu D. Richter, Paul Barach, T. Berman, G. Ben-David, and Zvi Weinberger, "Extending the Boundaries of the Declaration of Helsinki: A Case Study of an Unethical Experiment in a Non-Medical Setting," *Journal of Medical Ethics* 27 (2001): 126–29.

74. Thomas Mann, *Magic Mountain* (1924), (English trans., Harmondsworth, 1960), 466.

75. Shevell and Evans (n. 55 above), 355; Pross (n. 6 above), 33; Peukert (n. 16 above), 240.

Part I: What Is a Human Experiment?

Using the Population Body to Protect the National Body

Germ Warfare Tests in the United Kingdom after World War II

Brian Balmer

During the cold war, the threat of a germ warfare attack against the British population generated sufficient concern for government scientists to perform a series of large-scale, open-air trials using simulant biological warfare (BW) agents to assess the threat. After World War II a series of sea trials using pathogenic organisms were conducted off the coast of Scotland and in the Bahamas. The goal was an antipersonnel biological bomb. However, the weapon was never made. Although there is no open record or reference to any cabinet-level decision to abandon such offensive biological warfare research, by the mid-1950s scientists, military personnel, and policymakers all regarded the U.K. program as defensive. Within this defensive regime, scientists launched a series of open-air tests to assess the nature of the threat to the British population. This consisted initially of large-scale spray trials across much of England and Wales using a surrogate chemical marker. By the early 1960s the trials had graduated to the use of living nonpathogenic organisms that were intended to simulate a germ warfare attack.

Behind these experiments lay a double set of fears linked to disease and nationhood. And although tied here to the cold war context, these anxieties run deeper, pervading the notion of biological warfare whenever and wherever it has been contemplated. As sociologist Jeanne Guillemin argues, "Biological weapons evoke not just the fear of their violent assault on the individual body's life processes . . . but also fears of attack on the body politic, on who we are as citizens of a particular nation."[1] It is in this sense that the germ warfare tests in the United Kingdom can be regarded as experiments that aimed to protect—but that also involved—two intimately linked types of bodies.[2] Yet unlike other cases of "using bodies" for research, human bodies were not the target or subject of the research per se. People's bodies were only the focus of these experiments insofar as their ultimate aim was to safeguard the population from a biological warfare attack.

This chapter provides an account of the series of simulated biological warfare trials and the policy context within which they were performed.[3] It closes with a consideration of the involvement of the population in the trials. I will argue that the recorded discussions of the trials adopt a "discourse of exclusion." Within this discourse it was easy for their designers to abnegate any responsibility for the tests. Once the simulants were declared as harmless and the aims of the experiments as defensive, then humans were excluded from further consideration. By way of an alternative, I propose that involvement of the population in the tests can be construed so as to implicate these people fully as participants in the trial procedures.

The most detailed sources for narrating a history of germ warfare are the minutes and documents of the various secret government committees concerned with defense policy and research. A large number, but by no means all, of these papers are now open at the Public Record Office, Kew, London, and there are sufficient documents released to piece together an outline of events. In addition, while it must always be borne in mind that these are official records, there is a remarkable amount of detail in the minutes of meetings. Enough disagreement and discussion has been recorded to provide an insight into what options were considered as alternatives to decisions and recommendations on germ warfare. Without details of these deliberations, such decisions might otherwise appear, in retrospect, to have been inevitable.

Decision Making and Biological Warfare

Biological warfare involves the deliberate use of living organisms—usually pathogenic (disease-causing) microorganisms—to cause harm to humans, other animals, or crops.[4] After the Second World War, several decision-making and advisory bodies were established to oversee developments in the area. Starting with the research itself, work on both chemical and biological warfare took place at Porton Down, Wiltshire, in two separate but related Ministry of Supply establishments: the Chemical Defence Experimental Establishment (CDEE), and the Microbiological Research Department (MRD).[5]

Within the Ministry of Supply, the program of research at MRD was overseen by the Biological Research Advisory Board (BRAB), which consisted largely of independent scientists. BRAB was established in 1946 to provide independent scientific advice on "biological problems with special reference to micro-biological research carried out in the Ministry of Supply and extra-murally."[6] It was accountable primarily to the Scientific Advisory Council of the Ministry of Supply but provided technical advice to the Ministry of Defence's Inter-Services Sub-Committee on Biological Warfare. In 1947 the latter committee was reconstituted as a subcommittee of the Defence Research Policy Committee (DRPC). The DRPC was one of the highest level scientific committees in government with a remit to balance priorities across almost the entire spectrum of defense research.[7] A range of committees from the DRPC to BRAB thus had increasingly specialist responsibilities for germ warfare research in the United Kingdom.

The Status of Biological Weapons after World War II

At the close of the Second World War, the British research program in biological warfare had been running for five years at Porton Down, the site of research into chemical warfare since 1916. The work had been guided by a policy intended to prepare Britain for retaliation in kind at short notice if Germany initiated biological warfare. Research on offensive and defensive aspects of germ warfare continued into the postwar years, fueled by a similar retaliatory policy and an Air Staff request for a biological bomb to be built by the mid-1950s.[8] Research on biological warfare was elevated to the highest priority at this time, with the DRPC recommending that "research on chemical and biological weapons should be given priority effectively equal to that given to the study of atomic energy."[9] The chiefs of staff, who had recently formulated a secret policy of preparedness to use weapons of mass destruction, also agreed with this DRPC recommendation.

The high priority assigned to biological warfare soon translated into the building of new and expensive research facilities at the Microbiological Research Department.[10] In addition to the agents investigated during the war, anthrax and botulinum toxin, the program was broadened in scope to consider pathogens responsible for diseases such as brucellosis, tularemia, and plague. Another aspect of the burgeoning research program involved trials at sea using pathogenic organisms to ascertain the potential and limitations of a biological weapon attack. The first such trial, Operation Harness, was approved and then carried out off Antigua in 1948. In this operation, animals on floating dinghies were exposed to bacteria released upwind from sprays and munitions.[11] Similar trials, using an increasing diversity of pathogenic agents, took place between 1952 and 1955 off the coast of Scotland and off the Bahamas.[12]

As the trials expanded, however, biological weapons started to decline in their status as potential weapons of mass destruction. I have discussed the complex shift of biological weapons policy from an offensive to a defensive regime elsewhere.[13] In the context of the present discussion, it is sufficient to note several key factors that appear to have contributed to this gradual change. Practical demonstrations of the power of atomic weapons, the explosion of the first Soviet atomic bomb in 1949 and the British equivalent in 1952, marked a gradual overshadowing of Porton's patchy and tentative progress toward an antipersonnel biological bomb. In addition, from 1955 on, economic considerations came to the fore, and the continuation of *any* chemical and biological warfare research program was even questioned at several junctures.

No cabinet-level decision to abandon an offensive biological warfare program appears to have ever been made. Nonetheless, a close reading of the open statements on policy amongst various advisory and policy-making committees reveals an uneven gravitation toward a defensive posture. The 1955 DRPC review of defense research reported that biological research was now "mainly defensively aimed."[14] In the same year BRAB, the MRD's advisory board, discussed the establishment's future research plans "in the light of a policy directive by the Minister of Defence, which lowered the general position of Biological Warfare in the research and development programme."[15] By December the DRPC had "recommended that BW research should be restricted to that required for defensive measures. The work of the Microbiological Research Department would be substantially unchanged."[16] This declaration referred mainly to the basic research being conducted in MRD laboratories. Despite this announcement, BRAB advisors openly acknowledged that the new policy would have an effect on the continuation of field trials. Indeed, trials planned for the end of 1955 had already been postponed. The reason given made no mention of defensive policy: "The mass of data accumulated needed to be correlated with laboratory results and there was much stocktaking to be done before arrangements could usefully be made for further trials."[17] By the end of the following year the Admiralty had stated that the ship used in the trials, HMS *Ben Lomond*, would not be made available for future BW field trials.[18] The postponement had been changed into an indefinite deferral.

The Large Area Concept as a New Threat

In June 1957 a new Offensive Evaluation Committee, previously the Offensive Equipment Committee, held its first meeting. Located within the Ministry of Supply, its primary concerns and lines of responsibility were in the field of chemical warfare, although the committee members were also interested in the potential of biological warfare. At the June meeting "consideration was given to proposals that direct attack by conventional weapons limits the effectiveness of BW and that clandestine, off-target methods fully utilizing the insidious nature of biological agents would possibly enable a single aircraft to attack effectively tens of thousands of square miles."[19]

Previous research in the United Kingdom had concentrated almost exclusively on bombs as both a goal and the potential threat.[20] The new menace envisaged by the Offensive Evaluation Committee entailed an airplane or ship spreading a line of pathogenic biological agent some miles away from an area and thus spreading a deadly cloud across an entire region. This so-called Large Area Concept, involving an off-target attack, had been discussed as a general possibility at recent Tripartite conferences held jointly between the United Kingdom, the United States, and Canada.[21] Now researchers from both the chemical and biological sections at Porton, in a detailed report to the Offensive Evaluation Committee, put forward the case that the United Kingdom was especially vulnerable to the new threat.[22] The report discussed possibilities and problems concerned with maintaining an aerosol cloud of organisms over a long distance, keeping the organisms viable and their means of dissemination effective. It concluded: "In general, the feasibility of effective attack of very large areas with

BW agents is far from proven, but evidence is available which would make it dangerous to assume that it is not possible."[23]

This recommendation carried with it an implicit call for further research. The threat was likely but still needed to be proven. Whether or not this was deliberately implied in order to rejuvenate the trials program, a proposal to investigate and evaluate the large-area threat would have aligned with the new defensively oriented regime in biological warfare policy. Within a short period of time, scientists at Porton had embarked on a renewed series of open-air trials, supported by the rationale of the Large Area Concept as a defense threat.

Fluorescent Trials

The first series of trials that related specifically to the Large Area Concept involved the use of fluorescent tracer particles to simulate a biological agent cloud. Two aircraft field trials using zinc cadmium sulfide particles as simulants took place in the United Kingdom in 1957.[24] Zinc cadmium sulfide was also being used in parallel tests in the United States (Operation Large Area Concept), where it had been selected and used in various trials since 1950. Several years later, in a U.S. National Research Council investigation of the tests, the choice of zinc cadmium sulfide was justified on a variety of grounds: "[It was selected] not just because of its detectable glow, but also its particle size. Its particle-size range, 0.5–3 μm, approximates that considered most effective in penetrating the lungs. Other properties . . . were its economic feasibility; its lack of toxicity to humans, animals and plants; its stability in the atmosphere and its dispersibility."[25]

The first tests in the United Kingdom involved the release of 300 pounds of the particles along a 300–mile line off the western coast, with samples taken at meteorological stations in England and Wales.[26] In the trial, an airplane flew at about a thousand feet above sea level on a path running west of Newcastle, beginning dispersal of fluorescent powder over the Irish Sea and then continuing southwesterly until it was just a few miles from Wexford in Ireland.[27] The investigators deemed the results of a second similar study better than the first, being more in accord with what they had predicted: "The northern edge of the cloud was shown to be sharply marked; no particles had been collected at a station estimated to be 30 miles outside that edge."[28] Samples were taken at fifty-six locations and also by three sampling aircraft. Their conclusion, "if the samplers gave a true picture," was that 28 million people would have received a dose of one hundred particles.[29] Volunteer trials in the United States had already indicated that such a dose would have been effective for the debilitating but rarely lethal tularemia and Q-Fever diseases.

The trials progressed but not without difficulties. Commenting on the recently renamed Microbiological Research Establishment's (MRE) Annual Report for 1956–58, David Henderson, the Superintendent at MRE, informed BRAB that "work in the field was very slow because of the unfavourable weather. Results in the open compared well with the laboratory but were far too few."[30] Additional problems existed. Biologists, the board members noted, had "not run parallel" with meteorologists. Despite Henderson's report, other BRAB members expressed their view that correlation between work at Porton (using a rotating drum and a test sphere) and field results were equally problematic. Finally, the investigators wanted to know more about the effects of light on the survival of microorganisms in order to demonstrate a threat from large-area coverage.

While these obstacles were being discussed, the advisors also acknowledged the impossibility of undertaking large-scale toxic trials. In particular, one member of BRAB, Dr. R.W. Pittman, noted that "it was not possible to carry out large scale trials in this country with pathogens, and probably not with simulants either." However, it was mentioned at the same meeting that potential trials could be discussed with the United States and Canada at the forthcoming Tripartite conference in Canada. The significance of outdoor trials for the biological warfare researchers was underlined when, at the close of this discussion, the chief scientist identified large-area coverage as the "essence of the BW programme."[31] The "essence" of the biological warfare program, however, was not universally acknowledged; and germ warfare continued to remain a low priority in defense policy. In 1958 the chiefs of staff had declared that "the strategic value of BW in present known forms is insignificant."[32]

Despite such disregard from the military, five more long-distance trials of fluorescent particles were carried out between October 1958 and August 1959.[33] A "proofing" trial to determine effects of a release at sea was also carried out in October 1959. Yet just a few months earlier, in July, BRAB had been informed that obtaining the resources for the trials should continue only on an informal basis, that the DRPC "would not allot a high priority to the work, and the official priority might in fact be less effective than the present loosely defined arrangements."[34]

BRAB continued to remonstrate to higher authorities about the lack of attention being paid to the tests and, in its 1959 annual report, noted the slow progress of aircraft trials in the United Kingdom.[35] The board also "reaffirmed its belief that large area attack with BW agents constituted a major threat to the country's defences, and its profound dissatisfaction with the low priority accorded to this threat."[36] Henderson raised this matter again in a separate report to the Scientific Advisory Council (SAC) to whom BRAB was directly accountable. He complained that evaluation of the threat from a large-area attack was hampered and added that Porton could only obtain an aircraft for trials from the Ministry of Aviation at full cost.[37] The effect was to produce a supportive statement in the SAC annual report: "We referred last year to experiments which had been carried out both in this country and the USA in the dispersion of toxic agents released from aircraft. Results in both countries had shown that particles so dispersed were deposited over very large areas. It is the view of our Biological Research Advisory Board, with which we concur, that these results indicate that that method of attack could constitute a major threat to the defences of this country. Further work to evaluate the magnitude of the threat has been seriously delayed by the low priority which has been given to it."[38] A report on the matter had also been forwarded to the chief scientist at the Ministry of Supply, Sir Owen Wansbrough-Jones, "who had expressed sympathy with MRE's view but reiterated his doubt that seeking higher priority would give better results than the present reliance on goodwill."[39] While Henderson admitted in response that the air force had their own priority program, he said that "the time had come for the DRPC to face squarely this issue of what he believed to be a massive threat."[40]

Despite these setbacks, the tests continued. Trials with tracer particles from aircraft eventually succeeded in covering "areas of the UK with not less than 1 million and up to 38 million inhabitants."[41] By July 1960 a total of twelve such trials had taken place; the last two had been from on board ships.[42] In these cases, releases of fluorescent particles were made "simulating a breathable BW cloud as regards particle size"—one in the English Channel (8 Nov. 1959) and the second in the Irish Sea (10 Nov. 1959).[43] Meteorological Office, Ministry of Supply, and U.S. Air Force staff worked together to sample the clouds at permanent sampling stations across

England and Wales. The experimenters detected no great loss of particles into the sea, and this was taken to demonstrate that an attack from off the coast was a potential threat. The report of these trials concluded that "subject to the availability of BW agents which could not suffer a substantial loss of infectivity during dispersion and night travel for some 10 hours, a biological attack, mounted from a ship at sea, against the UK would have been feasible, and is likely to have been effective on this occasion over a very substantial area of sea and land."[44]

Once the likelihood of a threat under these conditions had been demonstrated to the satisfaction of the scientists and their advisors, the Offensive Evaluation Committee decided that a "sufficient" number of long-range trials had now taken place and resources should be diverted to smaller scale studies of particle loss from clouds.[45] One notable trial of this type did not attempt to spray a large area per se but to target a city. This trial, using a spray of powdered zinc cadmium sulfide, took place over Norwich on 28 March 1963.[46]

The aim of the Norwich trial was to test whether the heat emitted from industrial and domestic services in the city would disrupt the flow of a large cloud and whether the aerosol material would be deposited on the ground locally in larger concentrations than in the open countryside. Norwich was deemed large enough to generate sufficient heat and was situated "in an area as free as possible from topographical irregularities."[47] The line source released was 62 miles in length with the center of the line 24 miles upwind of Norwich. Sampling took place at thirty locations in the city, "mostly in the yards of adjoining police buildings and in the gardens of private houses," and at ten locations in the surrounding countryside. The results of the trial were inconclusive, although further trials were planned. As with the previous run of trials, the Norwich study did not involve human bodies in any direct manner. It was nonetheless predicated on human activities. The heat generated from industrial and domestic activities was an integral part of the trial design and hypothesis.

The Trials Go Live

BRAB, as mentioned earlier, had explicitly dismissed the possibility of large-scale trials using live pathogens or simulants in the United Kingdom. The board's view did not persist, and a year after rejecting trials with living organisms, BRAB members returned to a discussion of U.K. trials, now in the context of slow progress that was being made in U.S. and Canadian field trials. These tests had aimed to monitor the airborne survival of microorganisms in relation to the Large Area Concept. During this debate, the joint secretary of the board, John Morton, suggested that a paper requested by BRAB at the previous meeting might, on delivery, constitute a preliminary estimate of the scope and type of trials that MRE could undertake in a similar vein. Board members noted that any such future trials would have to be performed at night to enable maximum survival of the simulants. Further debate then ensued about the difficulties of extrapolation and generalization. One advisor, Professor Arnold Ashley Miles, director of the Lister Institute, raised these points in terms of whether a simulant might not differ less from a pathogen than pathogens from each other. Henderson responded that "close simulation of any particular pathogen was of no importance provided the general behaviour was of the same kind."[48]

As this debate moved away from purely hypothetical matters, Henderson raised the distinct possibility of conducting trials of nonpathogenic organisms over populated areas "on a much larger scale than ever before in the UK." The chief scientist was less sanguine and responded that "in view of possible objections" the Ministry's desert ranges, such as Maralinga in Australia, should instead be proposed as possible sites for trials. The debate closed with Henderson remaining enthusiastic and saying that "successful simulant trials were an essential prelude to pathogen trials." BRAB then recommended firmly that priority should be accorded to simulant studies in both the laboratory and in large-scale field trials. They also took matters a step further and proposed that a costing of simulant and pathogen trials should be prepared for the board by MRE.[49]

The idea of pathogen trials resurfaced in 1960 during a discussion of a paper entitled "BW Potential" by the Offensive Evaluation Committee. Their minutes contain the ominous hope that the United Kingdom would eventually be in a position "to carry out large scale trials with pathogens." The report also flagged areas where scientists required more information to assess the magnitude of the large-area threat. These topics included the effects of antipersonnel biological warfare agents on domestic and wild animals, field aspects of the stability of clouds of microbiological organisms, and agent detection. The Edinburgh biochemist and committee chairman, Professor Reginald Fisher, pointed out that since "most was known about agents which were merely incapacitating, he considered that more lethal agents should be looked for." However, Morton pointed out that MRE had no wish to duplicate searches for agents taking place in America, adding that, in any case, the Establishment "had no directive for development of agents or weapons."[50]

As an interim measure, Pittman proposed that a single large-scale trial with nonpathogenic bacteria could tie together the biological, physical, and meteorological aspects of the problem of cloud dissemination. The committee again noted problems with extrapolation. There was insufficient information to correlate the number of fluorescent particles collected during trials with the number of living organisms that would be inhaled during a biological warfare attack. Nor could correlations between trials in bursting chambers and in the field necessarily be extrapolated from simulants to other organisms. In all of this discussion there was no mention of safety per se, but the committee noted that "it would be difficult to get political agreement to releasing a cloud of simulant bacteria over populated areas. Fairly large scale trials had already been carried out with nonpathogens."[51]

Linked to problems with extrapolation, the choice of a suitable simulant for the proposed tests was also open to negotiation. Field trials using aerosols at Porton had been "steadily reduced in magnitude" since testing of pathogens, testing over large areas, and steady meteorological conditions were not available at the site. Previous trials at Porton "with the simulant *Serratia marcesens* [have] not been satisfactory chiefly because of poor stability of laboratory prepared suspensions." They concluded that "a more stable source of simulant would overcome this problem but would not erase doubt as to the applicability to pathogens of the information obtained."[52]

Even a suitable test site seemed elusive, with the Australian range at Maralinga proving "unpromising" and Christmas Island being "likeliest but would be very expensive because of logistic difficulties." However, despite BRAB's previous recommendations, the search remained a preliminary investigation "as there was no present intention to conduct trials."[53]

Scientists' concerns about the Large Area Concept had also generated a search for suitable methods to disseminate agents and simulants. Preliminary testing of spray devices had been initiated in 1957 and carried out jointly by CDEE and the National Gas Turbine Establishment.[54] A number of trials with different designs of sprays, using killed bacteria took place at Cardington and, when this site was found to be polluted and unsuitable, at a nonoperational Royal Air Force station at Odiham, some 50 miles from Porton. By 1960, various trials had also taken place back at Porton with live spores of the nonpathogenic *Bacillus globigii* in the cloud. In terms of safety procedures, Henderson had informed BRAB that "the material had been tested for non-virulence and in view of the ample precedent for such releases within the range he had not thought it necessary to seek permission."[55] Here the previous uses of *B. globigii* at Porton appeared to have played a key role in its definition as "harmless."

Large Area Concept and Early Warning

At about the same time that live simulant trials were being proposed in scientific and advisory arenas, the Large Area Concept became quite specifically characterized by the same scientists and advisors as a problem related to defense. A report discussing prospects for an early warning of a germ warfare attack alerted readers: "It can now be accepted that the principle of 'large area coverage,' by the dispersal of bacteria from aircraft, guided missiles or sea-borne craft, is satisfactorily established. A BW attack by this means may or may not commend itself as a strategic weapon; as a threat from a potential enemy it must certainly be taken seriously."[56]

Elsewhere, early warning was taken to be essential if those attacked by large-area coverage of biological agents were to gain protection from gas masks.[57] Early warning experiments aimed to provide at least two hours respite in order get 85 percent of the population under protection. It was envisaged that "the ten million civilian respirators, at present in store, would be issued to those members of the public whose duties made it essential for them to be in the open."[58] Whereas BRAB had previously flagged the Large Area Concept as the "essence" of the BW research program, the same committee now dubbed the problem of early warning as "the outstanding BW defence problem."[59] The rationale for trials thus became one of assessing the threat through an evaluation of the dissemination *and* detection of agents.

So the urgency with which early warning of attack was required and the perceived threat of a large-area attack combined under the defensively oriented research program to provide impetus for further trials. The overall level of concern about large-area attack expressed by BRAB and the SAC eventually found its way into the deliberations of a subcommittee of the DRPC in 1962. This subcommittee had been charged with writing a paper combining the findings of an ad hoc panel chaired by Sir Alexander Todd with the results from two operational assessments of chemical and biological warfare prepared for the chiefs of staff. In the introduction to their report, the DRPC subcommittee justified a renewed interest in the potential of chemical and biological weapons. Their reasons included the increased effectiveness of chemical agents, the future potential of nonlethal incapacitating agents, and the vulnerability of the civilian population to these weapons. The report also noted, echoing the concerns voiced by BRAB, that a key reason for renewing attention in the area was "the realization that large populated areas could be subjected to clandestine BW attack by aircraft or ships operating at distances of many miles from the target area" and the severe difficulty of detecting such an attack.[60]

The DRPC finally recommended "a modest expansion of effort on the offensive aspects of BW and CW in tactical situations, and further examination of the strategic potentialities of BW."[61] These recommendations were approved at cabinet level in May 1963. The United Kingdom was to develop a limited chemical retaliatory capability, and £470,000,000 over five years was to be spent on research into biological warfare. The amount allocated to biological warfare research was split between research on biological agents and field trials.[62] A few days before Cabinet met to discuss these matters, the minister of defense had been supplied with a briefing note that attempted to provide ammunition against any potential opposition to the large-scale trials. The note informed the minister that "experience in the past has shown that such trials cannot be kept entirely secret. The fitting in dockyards of naval vessels with hutches for guinea pigs etc. invariably attracts excited notices in the press. The consequent Parliamentary questions are bothersome but need not be more since the purpose of the trials is basically defence."[63]

The proposed single large-scale trial with nonpathogenic organisms mentioned earlier had now become a series of trials for which preliminary work was already underway. A year before obtaining Cabinet approval for trials, BRAB had sanctioned the use of live nonpathogens in trials that might involve exposure of members of the public "subject to vigorous testing of every batch of material in animals."[64] Between 1962 and 1963 the aerobiology and early warning groups at Porton were integrated, by which time Porton was using live organisms "wherever possible" in cloud tracking and sampling fieldwork. Local trials, presumably at Porton, took place to assess the viability of airborne organisms and test detection methods, while plans were made to conduct full-scale trials using ships and aircraft.[65] These included a CDEE trial at Portsmouth to study the penetration of a biological weapon simulant into a ship.

Then in May 1963, following the Cabinet approval and allocation of resources for further biological warfare trials, Walter Cawood, the chief scientist at the War Office, recommended that local tests at Porton on early detection of biological warfare agents be followed by larger tests. According to the chief scientist, it was now "necessary to release living simulants over much greater distances to allow time for natural (viable) decay." In addition, investigators wanted to check if the dead or dying organisms could be detected using techniques that worked in the laboratory. Consequently, trials had "been planned to simulate more realistic BW attack from the sea and air at distances of up to 50 miles from the south coast."[66]

The scientists at Porton were informed that further approval by the secretary of state for war for use of living organisms in trials was not required, although the chief scientist would keep him informed of progress.[67] The chief scientist justified the trials before BRAB on the grounds that the release of "harmless micro-organisms presents no special hazards" and that routine release of organisms from brewing, sewage disposal, and agricultural operations happen "on a vast and frequently uncontrolled scale, without public comment" and involve "organisms which have, potentially, a greater health hazard than the strictly controlled trials referred to in this note." He also mentioned that harmless microorganisms had been used on Porton ranges for years to "simulate BW agents in field trials. They are no more harmful than the normal background material of the air and are at a very low concentration by the time they have cleared the range area."[68]

Field Trials with Simulants

A progress report on field trials undertaken in 1963 and 1964 noted that "it can fairly be said that it is now usually possible to conduct an early warning or viability trial with a close approach to technical perfection; imperfect knowledge of the structure of the atmosphere is the cause of most failures."[69] By this time ten trials had been carried out along the south coast of England, mainly at night, using a ship as a source vehicle. Initially, these trials had used dead, stained organisms.[70] Most of the remaining trials used a mixture of live *Escherichia coli* (162) and *B. globigii* spores, the latter being used as a tracer.[71] More details of experimental procedures are available for later trials and provide some indication of the extent to which simulants were used. Between October 1964 and May 1965, thirteen trials took place at night in Lyme Bay and Weymouth Bay on the south coast.[72] Experimenters released bacterial suspensions of *E. coli* (162) and *B. globigii* along a line between 5 and 20 nautical miles long and between 5 and 20 nautical miles from the shore. A cloud was generated by four spray heads spraying bacterial suspension at about 4 liters per minute for between 55 and 113 minutes; the cloud was tracked by simultaneously releasing balloons carrying radar reflectors. Sampling of the cloud was performed on land at distances of up to 37 nautical miles downwind as the simulated biological warfare attack spread across the southern coast.[73]

Scientists drew several conclusions from this series of trials.[74] Primarily, they concluded that *E. coli* survived better when airborne in large particles than in small ones. The later trials had also used a novel technique that involved using "microthreads" of a spider's web to hold the microorganisms and study their viability in a simulated airborne state.[75] At the same time that trials with sprays occurred, sets of *E. coli* held on microthreads were exposed to the atmosphere in both exposed and sheltered sites. The experimenters concluded that survival of *E. coli* on microthreads was better than for airborne *E. coli*. Even with this disparity, the scientists concluded that the microthread technique was still in better agreement with field results than laboratory methods for estimating survival. Finally, the experiments confirmed "that the viability of *E. coli* is influenced by the previous history of the air mass."[76]

Ships and People

Trials entered a third phase toward the end of the 1960s and into the 1970s when assessment of risk moved away from the entire population and became focused on naval crews.[77] Tests were initially carried out using bacteria held on microthreads in order to compare their decay rates in different parts of a ship.[78] As well as using *E. coli*, one series of tests also used T7 coliphage, described in the trial report as "a virus-like organism parasitic for certain strains of *E. coli* and harmless to man and animals."[79]

These trials graduated onto a larger scale and involved sailing a frigate, HMS *Andromeda*, through a cloud of *E. coli* and *B. globigii*. In December 1970, the ship was exposed on three successive nights to clouds released from around three miles away, remaining in the cloud for between two and a half and eight minutes. Dosages measured in the ship were used to estimate the dose that would have been inhaled by crew members: "These minimal estimates showed that . . . a man in the Engine Room received between 700 and 2000 cells, in the Operations Room between 7 and 29 cells and in the Senior Ratings Dining Hall between 40 and 200 cells."[80] The trial report also noted that if readers took the numbers in the Operations Room and Dining Hall to be comfortingly low, they should be reminded that the estimated infective dose for *Pasturella tularensis* [*sic*], the causative agent of tularemia, was between ten and fifty viable cells.[81]

Investigators then shifted their attention to the fate of bacterial cells on the clothing of crew after a ship had passed through a cloud of bacteria. It is only at this point that human subjects enter directly into the ambit of the trials. In reports of these tests, the details of the crew's movements around the ship were spelled out in some detail. Under the materials and methods section of the trial report it was noted that "all the naval personnel engaged in trials on the upper deck during the periods when the citadel was in use wore CB suits, hoods and over-boots and entered the ship only via the cleansing station. . . . Although the use of CB suits in biological operations is in accord with Fleet instructions this was done to ensure that men exposed to challenge clouds wore similar outer clothing which could be discarded when they re-entered the ship."[82]

The trial, on HMS *Achilles*, took place over four days in January 1973. On the fourth day, the vessel spent about an hour in the bacterial cloud. The trial investigators concluded that "a ship in the normal ventilation condition is vulnerable to biological operations, and showed that the outer clothing of men exposed on the upper deck and in the machinery spaces was invariably contaminated with microorganisms."[83]

From an indirect exposure of crews to clouds of bacteria, the scientists moved on to assess the direct contamination of individuals with microorganisms. Tests were carried out mainly at Porton, although a few took place on ships.[84] An initial series of tests in 1974 demonstrated how the spores of *B. globigii* could penetrate inner and outer layers of clothing and settle onto hair.

A second series of tests, under the code name Gondolier, took place in 1976 in Portsmouth, where "the trials subjects were all young males with about three months of basic naval training." In six batches of five people, the men were sprayed for about four and six minutes with *B. globigii*, "during which time the spray operator walked slowly to and fro across the wind at a distance of about 80 metres upwind of the test subjects." The air was sampled as they undressed and cleaned up, then (reclothed) they entered a control room for an hour: "During this time card games were played, and for two short periods, each of about 30 seconds, mild physical exercises were carried out."[85] The sampling from air and clothing confirmed that there was contamination of the outer clothing and from secondary aerosols that arose during undressing.[86] Similar trials continued for another year when, according to the open literature, the series was terminated. The reports of these trials—either at sea or on land—provide no indication of what the participants were told about the nature of these experiments.

Using Bodies?

The idea of the body as research material or a research tool suggests a deliberate attempt to create subjects or measuring devices with people or parts of people. Within the sociology and history of the body, such attempts could readily be construed as a means of simultaneously producing knowledge about the body while exercising power over the body.[87] The bodies in the open-air biological warfare trials are not so obliging for theorists. With the exception of the last few trials, people were not used as subjects. They were not the target or tested object of the experiments, nor was the success of the trials dependent on their presence in the field. The body of the population was not used as an investigated variable or measuring device; no physiological measurements were taken; and all estimates of dosage from being sprayed were extrapolated to the lungs. Is there *any* useful sense in which these trials can be construed as "using bodies"?

Recent currents of thinking in sociology of science may provide some conceptual tools to include the unwitting population who were sprayed in the germ warfare trials. Social studies of science have informed and expanded our notions of "what counts" as part and parcel of doing science. Actor-network theory has notoriously extended our understanding of successful experimentation to include serious consideration of a heterogeneous array of human and nonhuman actants.[88] A related line of thinking has been pursued from the "social worlds" perspective that examines knowledge as practice. Sociologists in this tradition focus on the role of mundane tools, reagents, instruments, and their ilk. Rather than taking their suitability for experimentation as self-evident, this genre of studies has examined how such tools become defined as "right" in particular circumstances.[89] As with actor-network theory, their scope is expansive, taking into the analysis "everything in the situation, broadly conceived."[90] Everything includes workplaces, scientists, other workers, theories, models, other representational entities, research materials, instruments, technologies, skills and techniques, work organization, sponsorship and its organization, regulatory groups, and the audiences and consumers (intended and unintended) of the work.[91]

The people who were sprayed with zinc cadmium sulfide and bacteria do not fall easily into this list of categories. They were not part of a network or situation. They approximate an artifact needing to be excluded from consideration. At best, then, the population constituted what Casper and Clarke have called "implicated actors."[92] They were caught up in the experiment but were not part of it in any obvious manner. As such, these actors existed in the background, "between the categories, yet in relationship to them."[93]

The population can be viewed as "implicated actors" in several senses. First, the people who were sprayed were "implicated" in the obvious sense that they simply could not avoid being a part of the trials.[94] Secrecy and the scale of the operations combined to ensure that large numbers of people were exposed to simulants. They were also implicated in a less direct manner. Although the tests were performed under the auspices of a defensive policy, the Large Area Concept also marked a shift in the potential targets that would be countenanced in a biological weapons attack. The antipersonnel bomb that had been requested by the Air Staff immediately after the Second World War was intended for industrial or military targets. In contrast, while no biological weapon could ever be described as precise, the Large Area Concept envisaged an utterly indiscriminate weapon of mass destruction.[95] So the whole rationale underpinning the U.K. trials was to protect the very population who were being sprayed. Put bluntly, from the government and military perspective, it was "for their own good" that civilians were subjected to the trials.

Yet running counter to the inclusive thrust of these observations is a whole "discourse of exclusion." This entered at the point when scientists designated and selected harmless organisms for use in the trials. There is evidence, albeit scant, that the expert advisors considered the issue of safety and the "vigorous testing of every batch of material in animals."[96] Once harmless bacteria (and for that matter the inert particles) were chosen, then the people who would be sprayed with them shifted to the background and were no longer included as part of the trial design. The tests could proceed without reference to the population or their activities.[97] Yet as implicated actors, the move to approve particular nonpathogenic organisms could equally be construed as an influence by the population on the design of the trial. The choice of nonpathogenic bacteria could only be made with reference to the safety of the population—safety with respect to the bacteria, for example, when the chief scientist stated that the use of "harmless micro-organisms presents no special hazards";[98] and safety with respect to national security where "the consequent Parliamentary questions are bothersome but need not be more since the purpose of the trials is basically defence."[99]

Does this amount to anything of significance that could not be said in plainer terms? This depends on whether or not the boundaries of the trial, which, from the perspective of the experimenters excluded the very population who were ultimately to be protected, were coincident with moral boundaries of responsibility. Certainly, at the time, the discourse of exclusion appears to have been readily adopted; and since the organisms were deemed harmless and the objectives defensive, the main concern was not the population but instead appears to have been the potential political embarrassment arising if the tests were discovered.[100] These two pillars supporting the discourse of exclusion—safety and defense—can be further scrutinized.

First, the U.K. government had indeed shifted to a defensive policy on biological warfare at the time of the tests.[101] So even granted the thin line between defensive and offensive research, the declaration that these tests were defensive could be taken at face value. This does not render the experiments unproblematic, however, since a defensive national policy did not prevent the involvement of the experimenters in part of a wider research network. The Large Area Concept was explored collaboratively at Tripartite conferences while the United States was still in pursuit of an offensive biological warfare policy.[102] Whatever defensive intentions there were in Britain, the progress of parallel American and Canadian trials was discussed by British scientists; and results of the British trials would almost certainly have been made available to the offensively oriented U.S. program.[103]

Secondly, questions about safety are only apparently straightforward. Micro-organisms clearly do not come labeled as harmful or harmless. For example, *Bacillus subtilis* (or *globigii*) is not a pathogen, but it has been found with other *Bacilli* species in the cerebrospinal fluid of meningitis patients;

it has also caused respiratory problems for some factory workers coming into contact with enzymes from the bacteria when they were used in detergents.[104] On the other hand, a recent scientific review of the south coast trials concluded that the possibility of the bacteria causing a health hazard was very remote.[105] A further study by Dorset Health Authority, while admitting the uncertainties of a post hoc evaluation, also concluded that there was no evidence of a disease cluster, particularly a cluster that could be attributed to the germ warfare tests.[106] Even conceding the findings of these assessments does not close the matter of the role that the population played in these trials. Questions of safety, whether posed at the time or in contemporary discussion of the tests, are important but confine attention to the possible direct consequences of these trials.

Regardless of the physical risks involved, there is another set of questions about informed consent that any self-respecting bioethicist would ask.[107] It is a moot point as to whether or not it is anachronistic to raise these issues, although they have been asked of other events in recent history of science and medicine.[108] Under the discourse of exclusion, arguments about consent could be countered readily: the population were not part of the trial and therefore no consent was necessary, particularly given that an argument for secrecy could be made on security grounds.

However, raising the question of consent revisits my original questions about what it means to be "using bodies." I have pointed out that the whole rationale of the experiments was to protect the national population. In this respect, the locations of the trials were chosen to simulate a genuine attack on the United Kingdom. The inhabitants of these areas affected the design of the trials (albeit rather loosely), and they were actually subjected to spraying. Crucially, the issue of consent shifts our attention from the relatively narrow question of physical risks to a consideration of political risk. These are the risks that produce winners and losers by "excluding citizens from meaningful control" over science.[109] When the population body is treated as an artifact cluttering up an experiment, then it has lost control. Surely, this final shift of attention opens up a separate discourse around consent in which the body involved, the population, was as much a part of the trial as the ships, the bacteria, and the scientists themselves?

NOTES

I am grateful to Adam Hedgecoe for conversations about this chapter and to Joe Cain for helpful and insightful comments on an earlier draft.

1. Jeanne Guillemin, *Anthrax: The Investigation of a Deadly Outbreak* (Berkeley, CA, 1999), 245.

2. In fact, an older argument from cultural anthropology suggests that in certain societies or groups within societies, fear about the stability of the boundaries around the national body is readily transferred into fear about the stability of boundaries around the individual body (and vice versa). See Mary Douglas, *Purity and Danger: An Analysis of the Concepts of Pollution and Taboo* (London, 1966); and Mary Douglas, *Natural Symbols: Explorations in Cosmology* (London, 1970).

3. Some of the descriptive passages on the early trials (until 1963) have been taken from Brian Balmer, *Britain and Biological Warfare: Expert Advice and Science Policy, 1930–65* (Basingstoke, 2001) and have been reproduced here with the kind permission of the publishers, Palgrave.

4. For more complete definitions and discussion on these definitions, see, for example: SIPRI, *The Problem of Chemical and Biological Warfare* (Stockholm, 1971); Malcolm Dando, *Biological Warfare in the 21st Century* (London, 1994); Erhard Geissler, ed., *Biological and Toxin Weapons Today* (Oxford, 1986); Raymond A. Zilinskas, ed., *Biological Warfare: Modern Offense and Defense* (Boulder, CO, 2000); Joshua Lederberg, ed., *Biological Weapons: Limiting the Threat* (Cambridge, MA, 1999); Susan Wright, ed., *Preventing a Biological Arms Race* (Cambridge, MA, 1990).

5. The Chemical Defence Experimental Station was renamed the Chemical Defence Experimental Establishment in 1948. The Microbiological Research Department was renamed the Microbiological Research Establishment in 1957.

6. Public Record Office (hereafter PRO), WO195/9087. AC9091/BRB1, BRAB Constitution and Terms of Reference (11 July 1946).

7. See Jon Agar and Brian Balmer, "British Scientists and the Cold War: The Defence Research Policy Committee and Information Networks 1945–1963," *Historical Studies in the Physical and Biological Sciences* 28 (1998): 209–52.

8. See Brian Balmer, "The Drift of Biological Weapons Policy in the UK, 1945–1965," *Journal of Strategic Studies* 20 (Dec. 1997): 115–45.

9. PRO, DEFE 10/18. DRP (47) 53. DRPC. Future Defence Policy. 1 May 1947.

10. See Gradon B. Carter, "Biological Warfare and Biological Defence in the United Kingdom 1940–1979," *RUSI Journal* 137/6 (Dec. 1992): 67–74; Gradon B. Carter, *Porton Down: 75 Years of Chemical and Biological Research* (London, 1992); and Gradon B. Carter, *Chemical and Biological Defence at Porton Down, 1916–2000* (London, 2000).

11. The agents used were *Bacillus anthracis, Brucella suis, Brucella abortus,* and *Franciscella tularensis.* See Gradon Carter and Brian Balmer, "Chemical and Biological Warfare and Defence, 1945–1990," in *Cold War, Hot Science: Applied Research in Britain's Defence Laboratories 1945–1990,* ed. Robert Bud and Philip Gummett (Amsterdam, 1999).

12. Operations Cauldron (1952) and Hesperus (1953) off Stornaway on the island of Lewis; Operations Ozone (1954) and Negation (1954–55) off the Bahamas. See Carter and Balmer (n. 11 above).

13. Balmer (n. 8 above).

14. PRO, DEFE 10/34. DRP/P(55)50 (Final). DRPC. Review of Defence Research and Development (23 January 1956).

15. PRO, WO195/13460. AC13646/BRB139. Report by the Chairman on the work of the Board during the year 1955 (7 November 1955).

16. PRO, WO188/670. AC13524/BRB.M35. BRAB 35th Meeting (5 December 1955).

17. PRO, WO195/13313. AC13317/M126. SAC Executive Officer's Report for the 126th Meeting of Council (14 June 1955).

18. PRO, DEFE 10/281. DRPS/M(56)28. Defence Research Policy Staff Meeting (2 November 1956).

19. PRO, WO195/14170. AC14176/CDB237. Chemical Defence Advisory Board. Annual Review of the work of the Board for 1957 (4 November 1957).

20. Spraying had been discounted several times in various assessments of germ warfare. The threat from sabotage operations using biological agents remained ubiquitous. See Balmer (n. 3 above).

21. The Tripartite conferences had their origins in World War II collaboration between the three nations. They were generally, though not universally, regarded as useful platform for discussion and exchange of ideas in chemical and biological warfare research. See Gradon Carter and Graham S. Pearson, "North Atlantic Chemical and Biological Research Collaboration, 1916–1995," *Journal of Strategic Studies* 19 (1996): 74–103.

22. PRO, WO195/14064. AC14069.OEC.176. Ptn/Tu.1208/2129/57. Offensive Evaluation Committee. Study of the Possible Attack of Large Areas with BW Agents (1957).

23. Ibid.

24. PRO, WO195/14405. AC14413/BRBM41. BRAB 41st Meeting (18th July 1958).

25. National Research Council, *Toxicologic Assessment of the Army's Zinc Cadmium Sulfide Dispersion Tests* (Washington, DC, 1997): 117. Although this quote gives some indication of the reasons for using zinc cadmium sulfide, it is still a post hoc justification and therefore needs to be treated with some caution.

26. PRO (n. 24 above).

27. PRO, WO195/14593. PN68. Long Distance Travel of Particulate Clouds (23 January 1959).

28. PRO, WO195/14334. AC134342/CDb249 OECM3. Offensive Evaluation Committee 3rd Meeting (17 April 1958).

29. PRO (n. 24 above).

30. Ibid.

31. Ibid.

32. PRO, DEFE 41/56. JP(58)65(Final). Chiefs of Staff Committee, Joint Planning Staff. Biological and Chemical Warfare. Report by the Joint Planning Staff (30 July 1958).

33. PRO, WO195/14905. AC14914/CDB271 OECM5. Offensive Evaluation Committee 5th Meeting (15 January 1960).

34. PRO, WO195/14745. AC14754/BRBM43. BRAB 43rd Meeting (28 July 1959).

35. PRO, WO195/14811. AC14820/BRB172. BRAB Report 1959 (13 November 1959).

36. Ibid.

37. PRO, WO195/14902. AC14911/BRBM46. BRAB 46th Meeting (1 March 1960).

38. PRO, WO195/14876. AC14855. SAC Report for the year 1959 (March 1960).

39. PRO, WO195/14813. AC14822/BRBM44. BRAB 44th Meeting (31 October 1959).

40. Ibid.

41. PRO, WO195/14995. SAC70/BRBM48. BRAB 48th Meeting (16 July 1960).

42. PRO, WO 195/15014. SAC89/CDB281 OECM6. Offensive Evaluation Committee 6th Meeting (15 July 1960).

43. PRO, WO195/15142. Porton Note 203. Large Area Coverage by Aerosol Clouds Generated at Sea (22 March 1961).

44. Ibid.

45. PRO (n. 42 above).

46. PRO, WO195/15813. Porton Field Trial Report No.610. The Penetration of Built-up Areas by Aerosols at Night (7 May 1964).

47. Ibid.

48. PRO (n. 34 above).

49. Ibid.

50. PRO (n. 42 above).

51. Ibid.

52. PRO, WO195/14636. AC14645/BRB162. Microbiological Field Trials at Porton (29 April 1959).

53. PRO (n. 39 above).

54. PRO, WO195/15111. MRE Report No. 25. The Performance of Large Output Fine Sprays From Fast Moving Aircraft (16 January 1961).

55. PRO (n. 41 above). *Bacillus globigii* is currently named *Bacillus subtilis*.

56. PRO, WO195/15164. Early Warning Devices for BW Defence: The Situation in April 1961.

57. PRO, WO195/15168. SAC243/BRBM49. BRAB 49th Meeting (15 April 1961).

58. PRO, WO195/15693. SAC768/M19. SAC 19th Meeting (5 November 1963).

59. PRO, WO195/15267. SAC342/BRB204. BRAB Report by the Chairman on the work of the Board during the year 1961 (7 November 1961).

60. PRO, DEFE10/490. 24144. DRP/P(62)33. DRPC. Chemical and Biological Warfare (10th May 1962). Todd was also the chair of the DRPC's civil equivalent: the Advisory Committee on Scientific Policy.

61. PRO, WO195/15491. Appendix to SAC566. Technical Supplement to the Report for the Year 1962.

62. PRO, CAB 131/28. D(63)3. Cabinet Defence Committee Meeting (3 May 1963).

63. PRO, DEFE 13/440. Cabinet Defence Committee. Biological and Chemical Warfare Policy (D.(63)14). Brief for the Minister of Defence (1 May 1963).

64. PRO (n. 61 above).

65. PRO, WO195/15608. MRE Report No. 29. MRE Annual Report 1962–63.

66. PRO, WO32/20457. Chemical and Biological Warfare Field Trials (28 May 1963).

67. PRO, WO195/15610. SAC685/BRBM54. BRAB 54th Meeting (13 May 1963).

68. PRO (n. 66 above).

69. PRO, WO195/15749. SAC823/BRB231. BRAB. Field Trials Progress July, 1963—January 1964 (18 February 1964).

70. PRO (n. 58 above).

71. The number (162) denotes a particular strain of the bacteria used at Porton.

72. Harvard-Sussex Program Information Bank, SPRU, University of Sussex (hereafter HSP). MRE. Field Trial Report No. 4. The Viability, Concentration and Immunological Properties of Airborne Bacteria Released from a Massive Line Source. K. P. Norris, G. J. Harper, J. E. S. Greenstreet (September 1968).

73. A further four trials took place off Lyme Bay between 3 February and 26 April 1966.

74. HSP Information Bank. MRE Field Trial Report No. 5. Comparison of the Viability of *Escherichia coli* in airborne particles and on microthreads exposed in the field (n.d.).

75. PRO, WO195/15819. SAC893/BRBM56. BRAB 56th Meeting (13 March 1964); PRO, WO195/15750. SAC824/BRB232. A Method for Studying the Viability of Micro-organisms of Any Particle Size Held in the "Airborne" State in Any Environment for Any Length of Time (14 February 1964).

76. HSP Information Bank (n. 74 above).

77. There is very little documentation in the public archives regarding these trials. At the time of writing, the sources cited in this section had been declassified and were awaiting acceptance by the PRO. Copies of reports had also been made available in the House of Commons Library and Dorset County Library.

78. MRE Field Trial Report No. 8. The Survival of Airborne Bacteria in Naval Vessels: Tests with E. coli. G. J. Harper, J. E. S. Greenstreet, and K. P. Norris (May 1969).

79. MRE Field Trial Report 10. The Survival of Micro-Organisms Inside Naval Vessels: Tests in the Machinery Compartments of the Leander Class Frigate HMS Phoebe. G. J. Harper and K. P. Norris (February 1970).

80. MRE Field Trial Report 11. The Penetration of an Airborne Simulant into HMS Andromeda. G. J. Harper, F. A. Dark, and J. E. S. Greenstreet (October 1971).

81. Currently named *Francisella tularensis.*

82. MRE Field Trial report 14. Ship Defence Against Biological Operations.

Navy Trial Varan. G. J. Harper, F. A. Dark, and J. E. S. Greenstreet and Errington, F. P. (January 1974).

83. Ibid.

84. MRE Field Trial No. 15. Decontamination and Cleansing in Biological Operations. G. J. Harper, J. E. S. Greenstreet, and F. P. Errington (August 1974).

85. MRE Field Trial Report No. 21. Studies in the Protection Training Unit, Phoenix NBCD School (U). Navy trial Gondolier. (July 1976).

86. The trial report also contained recommendations on clothing to be worn and procedures for undressing and decontamination in defense against a biological attack.

87. Michel Foucault, "Body/Power," in *Michel Foucault: Power/Knowledge*, ed. C. Gordon (Brighton, 1980); Bryan Turner, *The Body and Society* (Oxford, 1984).

88. Bruno Latour, *Science in Action: How to Follow Scientists and Engineers Through Society* (Milton Keynes, 1987); Bruno Latour, *We Have Never Been Modern* (Cambridge, MA, 1992); Michel Callon and Bruno Latour, "Don't Throw the Baby Out with the Bath School! A Reply to Collins and Yearley," in *Science as Practice and Culture*, ed. A. Pickering (Chicago and London, 1992): 343–68.

89. A. E. Clarke and J. H. Fujimura, *The Right Tools for the Job: At Work in the Twentieth Century Life Sciences* (Princeton, NJ, 1992).

90. A. E. Clarke and J. H. Fujimura, "What Tools? Which Jobs? Why Right?" in Clarke and Fujimura (n. 89 above), 5.

91. Clarke and Fujimura (n. 89 above).

92. M. J. Casper and A. E. Clarke, "Making the Pap Smear the 'Right Tool' for the Job," *Social Studies of Science* 28 (1998): 255–90. The term is used but not elaborated on.

93. S. L. Star, "Power, Technology and the Phenomenology of Conventions: On Being Allergic to Onions," in *A Sociology of Monsters: Essays on Power, Technology and Domination*, ed. J. Law (London, 1991), 26–56, 39.

94. This was certainly the case for the civilian population. It is less clear whether the military personnel were given any choice over their participation.

95. See Guillemin (n. 1 above), 177–78, for a related discussion.

96. PRO (n. 64 above).

97. Except in the case of the Norwich spray trial, where the activity of humans in generating heat from industrial and domestic sources was a key component of the experimental hypothesis.

98. PRO (n. 68 above).

99. PRO (n. 63 above).

100. When the *Sunday Telegraph* first revealed that the tests had taken place in early 1997, then-Defence Secretary Michael Portillo appears to have adopted a semblance of this discourse. He is quoted as writing that the simulants were "judged to present no risk to health" (A. Gilligan, "Revealed: MoD's Germ Warfare Tests on London," *Sunday Telegraph*, 2 Feb. 1997, 1).

101. Balmer (n. 8 above).

102. Susan Wright, "Evolution of Biological Warfare Policy: 1945–1990," in Wright (n. 4 above).

103. As mentioned, the Large Area Concept had been discussed at the Tripartite conferences prior to the tests.

104. G. L. Mandell, J. F. Bennett, and R. Dolin, *Principles and Practice of Infectious Diseases*, 4th ed. (London, 1995), 1890–94. Less anachronistically, the meningitis observation was recorded in the scientific literature in 1963, the allergic reaction of detergent workers in 1969. See also Leonard Cole, *Clouds of Secrecy: The Army's Germ Warfare Tests Over Populated Areas* (Savage, MD, 1988) for a similar argument about U.S. outdoor trials.

105. B. Spratt, *Independent Review of the Possible Health Hazards of the Large-Scale Release of Bacteria During the Dorset Defence Trials* (Oxford, 1999).

106. C. E. Stein, S. Bennett, S. Crook, and F. Maddison, "The Cluster That Never Was: Germ Warfare Experiments and Health Authority Reality in Dorset," *Journal of the Royal Statistical Society, Series A* 164 (2001): 23–27. The authors note that interpretation of their results—which compared observed and expected levels of miscarriages, still births, congenital malformations, and neurodevelopmental disabilities—is limited by "a dependence on the calculation of expected health events, which are not routinely recorded, inconsistencies in definitions of the health events in the literature, inaccuracy of exposure ascertainment up to 38 years after the event, the calculation of rates from small numbers of observations and the inclusion of multiple generations in the study" (26).

107. See, for example, H. Wigodsky and S. Keir Hoppe, "Humans as Research Subjects" in *Birth to Death: Science and Bioethics*, ed. D.C. Thomasma and K. Thomasine (Cambridge, 1996); J. E. Sieber, *Planning Ethically Responsible Research: A Guide for Students and Internal Review Boards* (London, 1992); R .E. Bulger, E. Heitman, and S. J. Reiser, eds., *The Ethical Dimensions of the Biological Sciences* (Cambridge, 1993); Bernard Barber, "Experimenting With Humans," *The Public Interest* 6 (Winter 1967): 91–102.

108. In particular, the Tuskegee syphilis experiment. See J. H. Jones, *Bad Blood: The Tuskegee Syphilis Experiment* (London, 1993).

109. Sheila Jasanoff, "Product, Process, or Programme: Three Cultures and the Regulation of Biotechnology," in *Resistance to New Technology: Nuclear Power, Information Technology and Biotechnology*, ed. M. Bauer (Cambridge, 1995): 311–31.

CHAPTER THREE

Whose Body? Which Disease?

Studying Malaria while Treating Neurosyphilis

Margaret Humphreys

In 1931 Sir Henry Hallett Dale, the British physiologist whose work on neurotransmitters made him justly famous, heard a stimulating paper at the Royal Society of Tropical Medicine and Hygiene. Its topic, the use of malaria as a treatment for neurosyphilis, was interesting enough, but what really roused him to comment during the discussion afterward was the splendid opportunity this therapy offered to experiment, with no moral qualms, on a disease in humans. Since the production of malaria was a therapeutic measure of benefit to the patient, it was unambiguously justifiable. So "the ordinary moral difficulty, the ordinary complications which beset the work of a clinician, the ordinary dilemma of choosing between the advantage of the patient and the obtaining of new knowledge was absent in a peculiar sense in this particular case." Dale noted with admiration that "the opportunity had been given . . . of studying malarial infection as an experimentally-produced condition, as an investigation into the natural history of the infection." There must have been at least a tinge of envy in his voice when he labeled this capability as unique, for he "doubted if any other such existed in medicine, and he thought that here was the brightest hope for experimental therapeutics."[1]

From the 1920s on, physicians treating neurosyphilis began routinely to inject malaria parasites into their patients with the hope that the resulting high fevers would kill the syphilis spirochete in the central nervous system. Neurosyphilis, the late state of syphilis in which peripheral neurological dysfunction was often accompanied by insanity and dementia, did not respond to the arsenicals used for the treatment of milder, earlier disease. The malaria produced was real malaria, and it offered researchers an opportunity to study this disease in a controlled setting. Even though malariatherapy was "state of the art" for neurosyphilis treatment in the 1920s and 1930s, it is not quite true that researchers on malaria approached their induced subjects totally devoid of those "ordinary moral difficult[ies]" of which Dale spoke. This chapter explores the ethical and scientific issues raised by just such a research project in the work of Mark Boyd, a Rockefeller Foundation malariologist stationed in north Florida during the 1930s and 1940s.

Research on malaria was not at all what Julius Wagner-Juaregg, the inventor of malariatherapy, had in mind. This Viennese psychiatrist had been seeking a cure for the insane stages of neurosyphilis since the late 1880s, and in 1917 he tried the injection of malaria parasites.[2] Most historians have, in turn, focused on the syphilis story, with varying opinions about the efficacy of fever therapy. For example, three epidemiologists published an essay in 1992 in the *Journal of the American Medical Association* that attempted a meta-analysis of articles on neurosyphilis treated with malaria. The authors concluded that malarial fevers did not produce long-term benefit to syphilis patients, although with the caveat that there were no case-controlled trials that addressed the question with the rigorous standards set by modern science.[3] Other scientists have disagreed, citing the consistency of improvement rates seen in multiple studies.[4] Historian Joel Braslow considered malariatherapy as a part of his broader work on the history of psychiatric treatment in the early twentieth century. He drew the interesting conclusion that once psychiatrists had a treatment for this particular form of madness, they started speaking of their patients with more humanity and compassion. The availability of an apparently successful therapy changed not only clinical behavior but also clinical attitude.[5] From the point of view of the history of syphilis or the history of psychiatry, the advent of malariatherapy was an important episode in the development of somatic therapeutics.

Most contemporary medical writers of the 1920s, 1930s, and 1940s saw

it that way as well. Their focus was on syphilis and on how well fever therapy alleviated the symptoms of advanced disease. Researchers attempted
other means of raising the body temperature, such as infusing typhoid vaccine or heating the body with various mechanical devices. Malaria was
merely one way to reach a fever high enough to roast the spirochete without killing the patient. Wagner-Juaregg, who won the Nobel Prize in 1927
for his discovery, had tried a variety of substances before hitting on malaria.
At first he used tuberculin, Robert Koch's wonder drug for tuberculosis
in 1890, until it was discredited as dangerous and ineffective. Wagner-
Juaregg tried various other means to induce fever, including the injection
of typhoid vaccine, streptococci, and staphylococci. None worked as well,
or as safely, as malaria.[6]

Malariatherapy, or fever therapy more generally, became something of
a therapeutic craze in the second quarter of the twentieth century. One
textbook on fever therapy technique warned in 1939, "Like any new form
of therapy, therapeutic fever has been tried indiscriminately in most of the
diseases of mankind. . . . The enthusiasm and will to believe of many workers has far outweighed their clinical acumen." One thing was clear, however. "The therapeutic efficacy of these methods in certain diseases, such
as neurosyphilis, has been established beyond any reasonable doubt."[7] Popular science writer Paul de Kruif believed malariatherapy had so revolu-
tionized the treatment of neurosyphilis that he accorded Wagner-Juaregg
the stature of a medical hero, putting him alongside the likes of Semmel-
weis and Banting. Creating "friendly fevers" in patients with neurosyphilis
offered new and exciting hope to these otherwise hopeless victims of neurological degeneration.[8] Most physicians used the *vivax* malaria parasite,
which causes significant fever spikes but is rarely fatal, rather than *falciparum*, which could cause coma, renal failure, and death.

During the 1920s and 1930s malariatherapy was extensively employed
both in Europe and in the United States. At St. Elizabeth's mental hospital in Washington, D.C., for example, 2,158 persons were treated with
malaria between 1922 and 1936. A physician at the state mental hospital
in Bolivar, Tennessee, surveyed multiple American institutions in the early
1930s and could report an analysis of 8,354 such cases. In England, 821 patients with advanced neurosyphilis had been given malaria by 1926, and by
the end of 1929 that number had reached 3,155. In general, about one
quarter to one third of the patients improved enough to go home, and another third showed improvement. Anywhere from 1 to 10 percent died

during the therapy, although determining who actually died of the therapy was difficult, since these patients were already significantly ill of a fatal disease when the malariatherapy began.[9] There can be no doubt that from 1920 to 1945, malariatherapy for neurosyphilis represented the best that medical science had to offer for the neurosyphilis patient.

Most patients who received malariatherapy did so within mental hospitals, although a few private patients were treated at home. The physicians managing their care were specialists in mental disorders, not in malaria. So in most cases, if research was done at all, the focus was on syphilis, its neurological sequelae, and its response to various therapies. In a few instances, however, there were malariologists associated with the process, and this allowed for research on an artificially created and hence tightly controllable experimental disease. In England, S. P. James, a specialist in tropical disease who had retired from the Indian Medical Service and advised the government on tropical disease issues, supervised such research.[10] In the United States, the Rockefeller Foundation funded Mark Boyd's malaria work in the South, and in 1931 he initiated malariatherapy at a Florida mental hospital with the direct aim of studying malaria, not syphilis.

There were two methods of artificially inducing malaria, and both had their drawbacks. For either method, one first needed to acquire a patient with active malaria parasites to be tapped from his or her bloodstream. That patient had to be willing to delay treatment long enough for the parasites to be harvested. Then the process could go in one of two ways. The malaria blood could be drawn up in a syringe and directly injected into the syphilis patient, or the malaria source could become grazing ground for deliberately applied mosquitoes. Then those mosquitoes could be kept alive in cages long enough for the malaria parasite to go through its mosquito life cycle, an incubation time of one to two weeks. When the mosquito was ready, it could be applied to the syphilis patient. Variations on these two schemes were also possible, as in preserving the blood specimen and transporting it on ice or storing it. The mosquitoes could likewise be put under hibernation conditions and then activated for later use. Once the syphilis patient demonstrated malaria symptoms and parasites in his or her bloodstream, the process could begin again. Some hospitals kept a single strain of malaria alive for years by successfully passing it through one patient to another.

A word about the parasite's life cycle is necessary to clarify the difference between these two techniques. Under natural conditions, the infected mosquito injects malaria sporozoites with the saliva she inserts when she bites. The sporozoites enter the human body and go through various transformations. After a period of time in the liver, the parasite then moves into a red cell cycle in which it invades the red blood cell, multiplies, ruptures the cell, and then finds new cells to attack. The parasite, now called a trophozoite, may continue in this cycle, or it can transform into a male or female gametocyte, the sexual form of the malaria plasmodium. The female mosquito then takes up gametocytes with her blood meal. They mate in the mosquito's stomach, burrow through the wall, go through some more stages, and emerge as sporozoites in her saliva, ready to start a cycle again. So while the mosquito injects sporozoites, the prominent form present in the directly transmitted blood of the malaria patient is the trophozoite. Not surprisingly, the malaria cases induced by these two means have different clinical courses.

There are advantages and disadvantages to each technique. Blood taken from a malaria patient and directly injected into another person will carry malaria but will also carry all the other blood-borne diseases the malaria patient has. While physicians were not worried about AIDS in those early days, they did fear the transmission of hepatitis as well as strains of syphilis spirochetes that might worsen the receiving patient's condition. For these reasons direct transfer of malaria blood in order to create fevers in syphilitic patients was banned in England and was questioned elsewhere. What the technique had going for it was simplicity. It took very little training to learn to withdraw a few milliliters of blood from one arm and inject it into another's veins. The blood specimen could be transported on ice across hundreds of miles and maintained in cool storage until needed. The symptoms of malaria came on quickly, since the parasite was being introduced mid-cycle, as it were, having already gone through its early cycles elsewhere. The fever spikes were not as long-lived though; and some physicians believed that to cure neurosyphilis, the patient had to endure ten, fifteen, or even twenty bouts of chills and fever. In spite of these drawbacks, the direct blood injection route was the dominant form of malaria-therapy induction in the United States and Europe.[11]

To induce malaria by laboratory-controlled mosquito bites was much more difficult, although the infection created mimicked natural conditions much more closely. Malariologists, who tended to oversee malariatherapy with mosquitoes, had to learn how to preserve mosquitoes rather than killing them. First, anophelines had to be gathered from the wild, then kept alive and encouraged to breed. Once a set of them had fed on a patient with malaria, the mosquitoes had to be nurtured at a temperature optimal for both parasite reproduction and mosquito livelihood. This proved much trickier than expected, and entire cohorts of mosquitoes could die if conditions were not right. Then the "loaded" mosquito had to be convinced to bite the neurosyphilis patient. The mosquitoes were controlled by trapping them in jars open at the ends and covered with gauze. The gauze side was applied to the recipient's skin, which was sometimes pre-warmed to increase its surface circulation and attractiveness for the mosquito. The mosquitoes fed on sugar solutions to keep them alive between blood feedings; and the laboratory personnel had to balance sufficient nutrition with sufficient hunger, so that the mosquito lived but was also interested in taking a blood meal when offered. This process was further complicated by uncertainty about the infectiousness of the mosquito. Only after the mosquitoes had been applied to the patient were they dissected for evidence of sporozoites in their salivary ducts. If they had fresh blood in their gastrointestinal tracts and if there were sporozoites in their salivary glands, then it could be said with probability that the patient had received a dose of malaria parasites. This was all much more trouble than just injecting a small quantity of blood that was known to contain living organisms.

The positive aspects of this mode of malaria transmission were multiple, however. Since only the malaria parasite made it from the mosquito's stomach to her salivary glands, other contaminants of the donor blood were left behind. The induced malaria case was completely analogous to "wild" cases incurred outside of laboratory conditions, since the parasite entered the bloodstream in the sporozoite form and then proceeded through all of the normal human life cycle stages. So these patients presented a wonderful opportunity for studying malaria in closely controlled conditions. In particular, the time of infection could be determined with precision, whereas in nature the situation was much more confusing, since the patient might be exposed repeatedly to mosquito bites. The resulting infections could involve two or more different parasites or strains. Or, after multiple episodes of disease, the host and parasite could come into the sickly equilibrium of chronic malaria.

The malaria research community in the United States was rather slow to take advantage of this opportunity. While physicians in England had already infected thousands of neurosyphilis patients by means of "loaded" mosquitoes by 1930, no Americans were engaged in similar practices. Malariatherapy flourished, but it was of the direct-injection sort, which made the resultant cases of little use to the researcher interested in malaria and not syphilis. The Rockefeller Foundation decided to remedy this defect; and in the spring of 1931, it sent malariologist Mark Boyd (1889–1968) to Tallahassee, Florida, to open a malaria research station. Boyd had been employed by the Rockefeller Foundation since the early years of Rockefeller-funded research on malaria in the American South after World War I. Aside from occasional research trips to other malarious areas such as Brazil, Jamaica, and Sardinia, he had been working as a malariologist in the United States since then.

Boyd carried out his research in Tallahassee in cooperation with the Florida State Board of Health. Wherever he was stationed, Boyd sent quarterly diaries to his superiors at the Rockefeller Foundation in New York City. These accounts have personal detail (such as the death of a relative and his trip to the funeral) as well as information on the difficult research problems facing Boyd. This was a standard format for Rockefeller medicine men when away from home base; it was the way the foundation tracked its workers and kept abreast of their achievements and setbacks.[12]

The Boyd diaries offer an interesting window into the ethical scruples of one researcher in the early 1930s. These were the same years when physicians in the U.S. Public Health Service decided to continue a syphilis study carried out in Tuskegee, Alabama, as an "observation only" project with no apparent moral qualms at all.[13] Boyd, on the other hand, was overtly anxious about the health of his malaria patients, both those who served as donors and those who received the parasites. His donor cases were paid volunteers who cooperated on the assumption that delaying treatment would not significantly damage their health. Boyd may have been marginally less concerned about the syphilis patients, but still he appears to have been more worried about the immediate effects of malaria in them than were their treating physicians at the mental hospital. While Boyd had made a career out of preventing and treating malaria, he was new to the field of syphilis therapy. It may have come naturally to him to value treating malaria above treating syphilis, even while he was eager to study the natural course of malaria.

Boyd's first challenge was to get the superintendent of the nearby state mental hospital in Chattahoochee to cooperate with his plans for malariatherapy. The superintendent was enthusiastic and willingly offered up his many neurosyphilis patients for treatment. The hospital had both black and white patients, housed in wards segregated by race and sex. Since malariatherapy was a much-sought-after treatment in the early 1930s, it is not surprising that Boyd chose white patients for his preliminary experiments.

Boyd's position within the hospital's hierarchy was a curious one. He was not the physician who attended the syphilis patients and was only in an advisory or consultant role regarding their care. He or his assistant might apply the mosquitoes, but they did so under the direction of the state hospital physicians. Boyd himself actually spent very little time at the hospital and instead sent his assistant and co-worker, Warren Stratman-Thomas, to do the relevant work there. When controversy arose around issues of patient care, Boyd and Stratman-Thomas had to yield to the hospital's physicians. While they were malaria subjects to Boyd and Stratman-Thomas, they were first and foremost syphilis patients to the hospital's physicians.[14]

Having established a working relationship with the mental hospital and hence acquired experimental/therapeutic subjects, Boyd's next task was to find a source of malaria parasites. This proved an unexpectedly difficult task. North Florida was notoriously malaria-ridden in the 1930s, but Boyd needed not just any malaria patient. His human parasite reservoir had to meet several criteria: (1) have *vivax* (and definitely not *falciparum*) malaria; (2) have a case that had as yet received no treatment; (3) be willing and able to forego treatment for some period of time so that parasites could be harvested from the blood stream; and (4) be willing to have his or her blood drawn. Even the poorest of Florida's impoverished population could usually find a few dimes to buy a bottle of chill tonic. Although the quinine content of such patent medicines was low, it was enough to suppress the parasites and make acquisition of them from the patient's bloodstream unreliable. Finding a virgin case would occupy Boyd and Stratman-Thomas for much of May 1931.[15]

Boyd had a separate task to keep him busy during that summer. He had to acquire mosquitoes and build a housing unit for their preservation and propagation. This too proved much more challenging than he had anticipated. Upon first reading a diary entry dated July 14, 1931, one might be puzzled. It mentioned Cain, one of the research center's employees: "Cain conferred with sheriff of Jefferson Co., re theft of our pig. Interview not productive." It was not immediately clear why Boyd's group was keeping a pig or why they kept buying new ones to replace those stolen. It turned out that the pig had an important job. It was housed in a remote swampy corner of the county and served as bait to attract *anopheles* mosquitoes. Once captured, the mosquitoes were transferred to the insectary. But they could not keep a man out there all the time watching that pig, and no matter what pen and padlock were devised, some of the locals managed to bring home the bacon.[16]

The insectary also gave Boyd a lot of trouble. His initial incubator would not maintain the proper temperature. In early 1932 a much fancier model finally arrived after he had begged funding for it from the home office. That incubator proved a disappointment as well. It broke down almost immediately, and no local workmen could be found who knew how to fix such a sophisticated piece of equipment. There was a high learning curve in terms of the acquisition of information on how best to raise a mosquito to become a malaria transmitter. Boyd found, for example, that the optimal temperature for developing parasites was also one that promoted mosquito mortality. The malaria parasite could kill the mosquito if it multiplied too rapidly, it appeared. Feeding the mosquitoes offered another challenge. At one point he resorted to letting uninfected insects feed on the arm of the laboratory technician, a practice that one colleague denounced as inhumane.[17] In spite of all of these difficulties, Boyd was able to keep enough mosquitoes viable to carry out a series of malaria inoculations at the state mental hospital.

In early May, Boyd commenced giving malariatherapy by taking a young woman named Mabel over to the mental hospital, extracting some of her parasite-rich blood, and injecting 2.5 cc. into a woman with syphilis. Then he gave Mabel ten dollars and arranged for her to be given quinine. It is unclear why Boyd did not let mosquitoes feed on Mabel in order to start the mosquito-transmitted therapy—perhaps she was too ill, or perhaps she was unwilling to be made use of in such a way. In any event, his next malaria case was an 18–year-old white boy, whose parents brought him into Tallahassee to be fed on by mosquitoes. Boyd was very pleased with this source of parasites, calling him an "ideal subject."[18]

There were problems from the start, however. The boy had a severe outbreak of fever blisters on his face, which worsened with each malarious fever spike. Boyd had a local physician evaluate him to see if it would be safe to withhold treatment for a few days, especially since his blood was so rich in parasites. The local doctor said they could go ahead, and the mosquitoes were applied. The next day Boyd's anxiety continued, and he again asked the local physician to judge whether the boy could endure another chill. Again the doctor said it was all right, and more mosquitoes feasted on the boy's blood. Although Boyd had planned on giving the boy quinine after another day of mosquito feeding, he could not stand by while the adolescent's illness continued unabated. Boyd brought him some warm clothing and gave him quinine one day early. In the same week Boyd evaluated another malaria case but found her to be comatose. He not only advised immediate treatment, but he took quinine out to the woman himself. Clearly, it was hard for him, a malariologist, to stand by and not treat cases of malaria.[19]

Boyd was also very concerned about the initial syphilis patients who were treated with malaria. He felt frustrated due to his lack of control over their care once the parasites had passed into their bloodstreams. He sent Stratman-Thomas to make daily rounds on the malaria patients at the mental hospital. When one early case developed pus in his urine, Boyd wanted to at least "interrupt" the case with a brief dose of quinine. The psychiatrists at the hospital, who were trying to give the syphilis patients as many fever spikes as possible to control their underlying disease, did not agree, and the quinine was not given. Their concern was that the syphilis patients should receive the strongest possible dose of the curative syphilis therapy. A few days later another induced malaria case became dangerously ill. Boyd recorded in his diary: "S[tratman-] T[homas] reports that Mrs. C___ had several convulsions this morning, though he quotes Dr. Cobb as expressing the opinion that her condition is not serious and ascribes these symptoms to her paresis. I instructed S. T. to call hospital and order to have her given 30 gr. Quinine today. Also emphasized to S. T. that we must not permit any chances to be taken with the inoculated cases. Said that we could not under any circumstances permit an induced infection to continue unchecked if it in any way jeopardized the patient."[20] This patient did receive quinine, although not until more days had passed, so perhaps the psychiatrists offered some resistance. Another case was interrupted a few days later because his condition worried Boyd as well.[21]

Bad luck continued to dog Boyd in these early trials of malariatherapy. Another syphilis patient became seriously ill after receiving malaria. He had come into the hospital completely demented, signifying advanced neurosyphilis. During the malariatherapy the patient became very ill, perhaps because of an abscess on his thumb. When Boyd saw how sick the patient was, he ordered intravenous quinine. "Got in touch with Dr. Robertson, staff member on temporary duty, who gave 5 gr. quinine bichloride intravenously. Patient died within 5 minutes." Although Boyd was distressed by the death, he was also angry, since it incriminated the malariatherapy unfairly. This episode opened Boyd to the charge that malaria had killed the patient, even though he attributed the death to septicemia from the man's thumb infection. He, in turn, blamed the hospital's physicians, who had heretofore selected the patients for malariatherapy. This would have to stop. "In the past we have infected cases assigned to us by the hospital staff, . . . [leaving us to act] as a nurse would in applying prescribed treatment. In the future we will advise against the inoculation of any who are not in good physical condition." Although a few more deaths as well as "interruptions" of malariatherapy are recorded in the diaries, after the summer of 1931 the process seems to have gone much more smoothly.[22] Still, when Boyd and Stratman-Thomas described their technique in a 1933 article, they concluded with a cautionary note: "Although malaria therapy is most beneficial in many cases, it must be regarded as a heroic form of medication and should be employed with discretion."[23]

Boyd and Stratman-Thomas published several papers in which they described the research they did on *vivax* malaria through the medium of the mental hospital clientele. They were interested in immunity to malaria, so they studied what happened when patients were reinocculated with the same strain as they had been given initially. They found that the reinocculated patients (and not the controls who were exposed to the strain for the first time) remained free of disease, thus proving the existence of at least short-term immunity to particular *vivax* malaria strains. If, however, the patients who had received one strain were challenged with a second, they responded just as vigorously as if they had never had malaria. So Boyd and Stratman-Thomas were able to show that immunity to *vivax* malaria was strain-specific.[24] In a second article they reported on attempts to determine what quantity of mosquito bites was necessary to induce a case. They had found that mosquitoes varied in the number of sporozoites carried (and hence injected). The rate and quality of infection depended directly on the density of parasites in the mosquito, not on the number of mosquitoes applied or the number of bites.[25] Such information helped explain why some mosquito bites did not "take," but it had little practical application beyond the ranks of those attempting to induce malaria in syphilis patients.

Boyd and Stratman-Thomas had discovered something important, however, something that James in England was unlikely ever to see. They had black patients with neurosyphilis in their hospital population, and had a "devil of a time" giving them *vivax* malaria. At first they thought the mosquitoes were "bad" in some way—not full of parasites, or not biting properly, or held at the wrong temperature, or something. But mosquitoes from that same batch had no trouble infecting whites. In the fourteen months following June 1931, they had inoculated seventy-seven white patients with *vivax* malaria, and eight black patients. Of the eight, only three had any symptoms at all, in spite of large doses of sporozoites. "In the three [N]egro patients who were successfully inoculated, the clinical course of the infection was of exceptional mildness, so that little therapeutic benefit was to be expected," reported Boyd and Stratman-Thomas. Two of these patients had mild malaria attacks that lasted less than a week; the third showed symptoms eighty-five days after inoculation and then had fevers only to 100°F.[26] Boyd and Stratman-Thomas had discovered the innate African American immunity to *vivax* malaria.

The possibility that African Americans were less susceptible to malaria than whites had been suspected since the early years of the African slave trade. During the first centuries of Pan-American settlement, malaria limited the exploitation of the rich soils of the tropical and subtropical zones. While *vivax* malaria had come to the New World from Europe, the Africans brought *falciparum* malaria in their unwillingly transported bodies. The result was that wherever whites and blacks mixed in the Tropics, severe malaria followed, mowing down Europeans while apparently sparing black people. Europeans were familiar with the fever spikes of *vivax*, but most had not seen the malicious and deadly malaria that came to characterize the American colonies in a coastal band from South Carolina, through the Caribbean, down the Mexican coast, and into Latin America. These colonies quickly acquired a reputation as unhealthy and feverish, spurring the development of the African slave trade. White workers avoided the sickly lands where possible; Africans were forcibly brought there, and at least managed to escape, by and large, from the ravages of the local fevers. Thus, when colonial and antebellum apologists for slavery argued that black people were particularly suited to labor on their hot, humid plantations, their argument had some basis in fact.[27]

Awareness of racial variability with regard to malaria served a strong social purpose in arguments supporting slavery. Given their tolerance of malaria, this line of reasoning went, blacks were destined by God and biology to labor under tropical conditions not suited to white people. After the U.S. Civil War, discussion of this phenomenon largely faded from the medical literature on race and on malaria. By 1900 when white physicians discussed "the Negro health problem," they were addressing the appalling mortality figures that characterized the black population, especially those living in cities. The principal culprits were tuberculosis and venereal disease, diseases made all the more disturbing because of the possibility of transmission to the white race: blacks were ubiquitous in white households, serving as maids, caring for children, and waiting on tables. Blacks were no longer seen as having a feature that made them healthier than whites, but rather as being particularly diseased and dangerous. Some even argued that the race had so degenerated since the years of kindly paternalistic care by slave owners that it was in danger of extinction.[28]

Furthermore, it was widely evident in the twentieth-century American South that blacks suffered disproportionately more from malaria than whites did. Over and over again, mortality statistics showed more blacks than whites dying of malaria, often in ratios as high as two to one. For example, a researcher for the U.S. Public Health Service, Kenneth Maxcy, found twice as many cases of splenomegaly (a sign of malaria) among black Mississippi delta school children as among white.[29] This difference was also borne out in studies that looked at parasite rates between the races. Although one can argue that these statistics have various biases, it remains clear that public health officials and the public at large saw the black population as more at risk for malaria than the white one. When the researchers compared black and white, the blacks usually equaled, if not much exceeded, the parasite rates of whites. Of course such percentages were very much dependent on the population surveyed, but it at least indicates that public health researchers had no problem finding abundant malaria infestation among southern blacks.[30]

So expectations about blacks and malaria in the early 1930s did not predict that it would be hard to give them malaria. They seemed to get it just fine out in the swampy southern world. The fact that most of those cases were caused by *falciparum* parasites was obscured by a variety of factors. Very few doctors in the American South were equipped with the knowledge, the microscope, or the inclination to make an accurate diagnosis. Many poor blacks never saw a doctor, preferring to dose themselves with patent medicines containing quinine if they sought any treatment at all.

Even when public health researchers did parasite surveys, going into schools or communities and getting blood smears from the population, the fact of racial differentiation by parasite was not recognized. Distinguishing the parasites under the microscope is not always easy. Furthermore, as Boyd and Stratman-Thomas showed, black people could have parasites in their bloodstreams but show no clinical signs of infection. So parasite surveys tended to overestimate the prevalence of *vivax* malaria in the black population, if it was measured at all.

It was not until 1975 that the mechanism of African American resistance to *vivax* malaria was sorted out. In that year a researcher showed that about 95 percent of sub-Saharan Africans have a characteristic of their red blood cells that causes them no apparent harm. Their red blood cells are missing a cell wall structure called the Duffy antigen. Without this structure, the *vivax* parasite apparently cannot gain entrance into the red blood cell. This represents an absolute immunity: bearers of "Duffy-negative" red blood cells will never have a case of *vivax* malaria, although under certain circumstances they may have the parasites swimming in their bloodstreams. They can, in other words, be infected but not sick.[31] This trait should not be confused with the myriad of genetic defects that protect many Africans from *falciparum* malaria. The most famous of these, the sickle-cell trait, makes it more likely a child will survive early *falciparum* infections and makes the resultant disease less hazardous, even in adults. This and other hemoglobin abnormalities help protect many blacks from *falciparum* malaria, but the protection is only partial, not absolute.[32]

This complex situation with regard to malaria immunity explains the variant disease mortality in the early years of New World colonization and conquest. African Americans were better able to survive *falciparum* infection than whites due to inherited genetic traits (as well as acquired immunity from a childhood spent exposed to malaria). While blacks gave deadly *falciparum* to whites, they were largely unscathed by the *vivax* malaria that whites carried to the New World. The role of sickle-cell trait and other hemoglobin variants in determining the malaria death rate was not demonstrated until the 1950s, so Boyd and Stratman-Thomas were the first to establish the existence of a racial resistance to malaria. Their work did not light any ideological fires, however, and was received quietly by the research community. Still, the fact of partial resistance paved the way for other researchers, who ultimately established the presence of genetically determined malaria resistance and immunity.[33]

Although Boyd recognized that his discovery about blacks and *vivax* malaria was significant, it did not help him in treating them for neurosyphilis. But it did set up an enticing opportunity. So far he had been careful to exclude *falciparum* parasites, knowing how dangerous they were. But there was much he wanted to learn about *falciparum*, so the fact that syphilitic blacks could not benefit from *vivax* and might benefit from malaria therapy, helped him justify the use of this heroic course. The relative immunity of African Americans to *falciparum* was unknown to him, although he would have recognized the existence of acquired immunity in adults who had grown up in a malarious area. He eagerly, perhaps too eagerly, embraced the opportunity to study the natural history of *falciparum* with the same degree of control as he had found with *vivax* malariatherapy patients. Accordingly, in late October 1931, he gave his first dose of *falciparum* to a black man suffering from neurosyphilis.[34]

Boyd was very anxious about the series of patients he injected with *falciparum* over the next couple of weeks. He ordered Stratman-Thomas to sleep over at the hospital so their care could be carefully monitored. After the initial rounds of fever went satisfactorily, Boyd relaxed his watchfulness. Then, on November 21, "S.T. called about 6 p.m. to say that had just received message from hospital that J——, one of our falciparum cases had had a temp. of 105°[F] for 4–5 hours and was then in a coma. T. said he asked Dr. Watson to give intravenous quinine immediately." Note again that the care of the malaria patients was only indirectly in the hands of Boyd and Stratman-Thomas. Boyd sent Stratman-Thomas out to Chattahoochee to examine the patient, where he found that the patient had died before the quinine could be administered. Later *falciparum* patients died as well, although Boyd often attributed their deaths to causes other than malaria.[35]

These deaths did not dissuade Boyd from further use of *falciparum*. By 1935 he had inoculated seventy-two black patients with this strain of malaria and reported on his results in the *American Journal of Tropical Medicine*.[36] Of this group sixty developed cases of *falciparum*, four died, and forty-nine had to have their fever bouts interrupted with quinine before the desired number of fever spikes had occurred. The criteria for termination were either parasite counts over 100,000 per cubic millimeter of blood inspected, or a fever above 104°F. (The more dense the parasites, the more severe the disease, the more likely death will ensue.) There were four white patients in the cohort receiving *falciparum* (chosen because they were refractory to *vivax*), although Boyd does not draw any conclusions regarding racial susceptibility since the white sample was so small. He found that one mosquito bite could be sufficient to cause a case of *falciparum* and that increasing the dosage of parasite did not effect the course of the disease. He also published extensive data on the duration of *falciparum*, the number of fever peaks, and the pattern of febrile episodes. Finally, Boyd noted that the *falciparum*-induced fever therapy was just as effective in neurosyphilis as that obtained from *vivax*.

In 1938 Boyd and colleagues published a paper summarizing their experience with patients at the Chattahoochee state mental hospital. Just over two hundred patients had been injected with malaria parasites, 75 percent via mosquitoes and 25 percent via direct intravenous inoculation, as well as being given the standard medical treatment with arsenicals. In addition to using *vivax* and *falciparum* malaria, Boyd had also injected some cases with quartan, or *malariae* malaria.[37] He found that these latter patients did well, but keeping the quartan strain alive and active was very difficult. Although one quarter of the malariatherapy patients were dead at the time Boyd was writing, he believed only 7 percent of deaths were due to malaria. In comparison, only 45 percent of the patients treated with medication alone were still alive. All told, in the malaria group, 31 percent were deemed "in remission," 23 percent "improved," and 19 percent "unimproved" (with 25% dead). "The mortality experience in the 2 races is similar," Boyd and colleagues reported. Since "in remission" meant, by definition, that the patient was well enough to go home, "the remission rate among colored persons was actually better than shown, as the furloughing of many colored patients who showed satisfactory improvement was impracticable for lack of guardians."[38] In summary, Boyd and colleagues concluded, "Malaria therapy combined with chemotherapy gives very much better results in the treatment of neurosyphilis than chemotherapy alone."[39]

So Boyd saw himself as providing a valuable therapeutic option to these patients, and hence he justified his use of them as experimental subjects. It is interesting to note that they were someone else's *patients* (the doctors at the mental hospital), and mainly for him *subjects* of his research. Still, he could not maintain the cool researcher objectivity that would be expected if he only saw them as research objects. He was quite worried about causing harm to either the sources of his parasites or to their recipients, although some of this concern may have arisen less out of compassion than out of concern for the good name of the Rockefeller Foundation. This issue came particularly to the forefront when private patients, not inmates of the state mental hospital, applied to Boyd's group for malariatherapy.[40] In January 1934, for example, a patient from Tampa showed up at the malaria research station for inoculation. "The usual release form was secured from patient and 3 mosquitoes fed."[41] Nowhere else did Boyd mention such a release or consent form, so it is not clear if he required it only for nonhospital patients or for everyone. Many of the hospital patients were too demented to make any decisions for themselves, and the acquisition of consent from relatives was only mentioned in regard to securing autopsies, not to giving malariatherapy.

The only other insight into Boyd's worries about liability came in this comment written in December 1932: "FFR received wire from JAF about liability of RF if we do inoculations of private paretics [neurosyphilis patients] in Ala. and Ga. FFR not inclined to view as desirable."[42] FFR was Frederick F. Russell, director of the International Health Division and one of Boyd's superiors within the Rockefeller Foundation (RF); JAF was John A. Ferrell, who had supervised the earlier malaria campaigns funded by the Rockefeller Foundation during the 1910s and 1920s, and so was considered an in-house expert on the disease.[43] Boyd, in other words, was being ordered not to treat private patients from Alabama and Georgia since that would put the Rockefeller Foundation at too much risk, not only strictly because of legal liability but also because of possible negative publicity.

One ethical issue not addressed directly in either Boyd's published papers or in the diaries was the possibility that hospital-induced malaria could spread to the surrounding community. Indeed, Stratman-Thomas himself acquired malaria while working in the mental hospital. Although the white wards were screened, the black wards were not made mosquito-proof until 1935.[44] The only mention of concern about this community hazard is a cryptic entry in Boyd's diary for November 20, 1932: "ST got an autopsy on a child dying from malarial hematuria living near Chattahoochee, we paying $15.00 toward burial."[45] Given the presence of bloody urine, a sign of renal failure, the child probably died of *falciparum* malaria. Did Boyd pay $15 toward her funeral costs only to have the opportunity to do an autopsy on a malaria case? That is possible, but there is nothing further recorded about the autopsy or about why the case might be particularly interesting. Or did he feel some guilt at the prospect that the infecting *falciparum* parasites had come from one of his malariatherapy patients and was, in effect, paying the parents off? I suspect the latter, but there is no further information about community malaria cases to support my conclusion.

Although the idea of giving malaria to cure another disease seems strange to us today, it was an accepted therapy for a devastating disease when Boyd took it up in the 1930s. In fact, it still has some appeal for patients with modern, incurable diseases, as evidenced by a Mexican clinic that offered malariatherapy for late-stage Lyme disease in 1989 and a recent Chinese trial of using malaria to treat AIDS.[46] Boyd's methods do make one uneasily aware of the slippery slope that might lead researchers to induce malaria in patients with no underlying disease in order to test medications or perform other research on malaria. This indeed happened during World War II. Conscientious objectors and prisoner volunteers in the United States agreed to receive malaria so that new medications could be tried out on them and older medications tested for appropriate dosage intervals and quantities. The Nazis took it one step further. When the need to control malaria among German troops stationed in the southern areas of Europe and Southwest Asia became acute, German doctors turned from testing malaria drugs on syphilis patients and volunteers to using inmates at concentration camps. The physician who had supervised these studies at Dachau was hanged in 1946 after his meticulously detailed records survived the camp's liberation. During his trial he defended himself by saying, "I admit that people had to suffer because of each experiment, mostly from depressions. Yet, the scientific interest to protect millions of people from this disease and to save them was predominant."[47]

There has been a recent analysis of the ethical aspects of using malariatherapy research data, although not in regard to Boyd's work. In 1999 the *American Journal of Tropical Medicine* published a series of papers by William Collins and Geoffrey Jeffrey that used retrospective analysis to describe the course of 474 subjects who received *falciparum* malaria at two mental hospitals in the American South between 1940 and 1963. All subjects were patients diagnosed with neurosyphilis. The authors' intention was that their data base of induced infections could provide essential information about the natural history of *falciparum* malaria to be used by researchers studying malaria vaccines.[48]

Accompanying the articles presenting this data was an essay entitled "Another Tuskegee?" by ethicist Charles Weijer of Dalhousie University.

He asked whether it was ethical to use the information that Collins and Jeffery presented. Similar questions have been asked, for example, about the use of Nazi medical research, since its subjects were unwilling prisoners who were frequently harmed or killed during the experiments. Weijer drew several conclusions about the malariatherapy research. First, he noted that 60 percent of the malariatherapy patients were black, and 40 percent were white. Hence, he felt that blacks were not selectively targeted for the research, unlike in the Tuskegee case. Second, he pointed out that the patients or their families gave consent for treatment within the hospital and that malariatherapy was a standard treatment of the day. Finally, he argued that since the subjects were clinical patients receiving appropriate care for their illness, they could not simultaneously be research subjects. He made the distinction based on whether the malariatherapy patients received extra interventions for research purposes that they would not otherwise have received, and he concluded that the answer was no. Altogether, he found that Collins and Jeffery "present data that will be invaluable to future malaria research . . . in an ethically supportable manner."[49]

One aspect of Weijer's analysis is based on a misunderstanding of the Collins and Jeffery data. They tell us that a total of 1,053 patients received malariatherapy and that roughly 60 percent of these patients were black. Yet their four papers all look at the subset of the 1,053 patients who received *falciparum* parasites—474 patients. We are not told the racial breakdown of this population. In total there were 635 black patients who received malaria of any kind. It is likely that all, or almost all, of the 474 *falciparum* patients were black, with most of the remaining 161 black patients receiving *Plasmodium malariae* (the parasite of quartan malaria). As a 1941 description of the malariatherapy program at the South Carolina State Hospital (one of the two analyzed by Collins and Jeffery) noted, "In Negroes tertian malaria [*vivax*] does not develop with any degree of success."[50] So Weijer missed the point that almost all of the patients who received the most dangerous of malarias were black, and he failed to analyze the ethical implications of this fact. He provided no guidance about the willingness of physicians, including Boyd, to give *falciparum* to black patients. Was it made easier because the recipients of this heroic therapy were slightly less valued as human beings than white patients? Perhaps, but there is no evidence here to support this conclusion. Certainly Boyd, for one, was acutely aware of the dangers of this potentially deadly parasite.

Whether his anxiety as a researcher was balanced by his conviction that this was best for the patient who had no other way to benefit from malaria-therapy is something that his diaries and papers do not reveal.

Weijer's distinction between research subjects and clinical patients is interesting and raises questions about Boyd's work. Did his patients receive any interventions that would not have otherwise happened because they were subjects of his study? Well, yes—they received malariatherapy. Florida State Hospital did not provide this sort of treatment at all prior to Boyd's arrival. Whether this was good or bad depends on the unanswered question about the treatment's efficacy. This aside, the issue is more complex in the Boyd case than in the one analyzed by Weijer. The patients described in Boyd's work had two distinct sets of practitioners providing care for them. One set, the doctors at the Florida State Hospital, had as their primary goal the treatment of syphilis, and they were willing to subject patients to dangerous levels of malaria in order to maximize the conquest of neurosyphilis. Boyd, on the other hand, had inherent conflicts in his goals. He wanted to gather data on the natural history of malaria, but at the same time, he wanted the patients to endure only safe levels of malaria, particularly since he himself had given it to them. If anything, the physicians directly responsible for the medical care of the Chattahoochee patients appear to have been more callous about their suffering (and accepting of potential death) than the researcher who might be assumed to be the more distant and uncaring participant.

One final note should be made about Boyd's peculiar difficulty of inducing malaria in his African American subjects. His paper on the innate immunities of blacks to *vivax* caused no splash at all. It is cited appropriately by later works on the Duffy antigen that explained the immunity, but otherwise it does not appear in discussions about race and health issues that took place during the 1930s. It was a discovery that served no social purpose in its time. Southerners were no longer claiming that somehow blacks were particularly suited to toil in tropical climates as enslaved labor, since this antebellum argument had no function once the Civil War had made the issue moot. Boyd's discovery came a century too late for slavery's defenders. Social reformers who promoted the cause of black Americans likewise had little use for the information about *vivax* immunity. Their line of rhetoric blamed the various health problems of the black race on socioeconomic oppression; they specifically opposed the idea that blacks were biologically different and hence more susceptible to, say, tuberculosis. Instead, they argued that blacks were the same as whites biologically and would be just as healthy if they lived in adequate housing, ate nutritious food, and worked in safe environments.[51] Again, they had no use for Boyd's data, even though it could be touted as showing that blacks were actually stronger and more fit than whites in one respect. Boyd's research met an ideological void.

Joel Braslow has argued that malariatherapy tended to make psychiatrists see their neurosyphilis patients more as people and less as dehumanized, demented creatures suitable only for control and warehousing.[52] But it is certainly possible to argue the opposite point of view, especially if the physician approaching the patient was primarily interested in malaria and only secondarily in the treatment of neurosyphilis. In this case the patient becomes a subject and becomes vulnerable to potential abuse. Mark Boyd and his colleagues struggled against this inclination, while at the same time they appreciated the experimental gold mine that the neurosyphilis patients offered. It was not accidental that their patients/subjects suffered from one of the vilest social diseases known and were generally condemned by society for having that illness in the first place. This may have helped create a patina of "otherness" that made the creation of a research mentality possible. Yet Boyd resisted this impulse, and he seems to have retained his role as a caring physician, torn in his desires both to study and to treat malaria in spite of working with patients that would have stretched any definition of attractive humanity. His use of *falciparum* allowed him to offer black patients both the dangers and the rewards of malariatherapy and to pursue research on an otherwise inaccessible disease. But in so doing he continually faced Dale's dilemma of "choosing between the advantage of the patient and the obtaining of new information."

NOTES

1. Sir Henry Hallett Dale, comments following S. P. James, "Some General Results of a Study of Induced Malaria in England," *Transactions of the Royal Society of Tropical Medicine and Hygiene* 24 (1931): 528.

2. [Julius] Wagner-Juaregg, "The Treatment of General Paresis by Inoculation of Malaria," *Journal of Nervous and Mental Disease* 55 (1922): 369–75; Magda Whitrow, "Wagner-Jauregg and Fever Therapy," *Medical History* 34 (1990): 294–310.

3. Stephanie C. Austin, Paul D. Stolley, and Tamar Lasky, "The History of

Malariotherapy for Neurosyphilis: Modern Parallels," *Journal of the American Medical Association* 268 (1992): 516–19.

4. Eli Chernin, "The Malariatherapy of Neurosyphilis," *Journal of Parasitology* 70 (1984): 611–17; Magda Whitrow, letter to the editor, *Journal of the American Medical Association* 270 (1994): 343; Henry J. Heimlich, letter to the editor, *Journal of the American Medical Association* 269 (1993): 211. There are enough immunological similarities between the syphilis and malaria organisms that the presence of malaria can generate a false positive result on syphilis testing. Such parallels could mean that malariatherapy did not act simply through temperature effects but also through stimulation of a specific immune reaction against the spirochete.

5. Joel T. Braslow, "The Influence of a Biological Therapy on Doctor's Narratives and Interrogations: The Case of General Paralysis of the Insane and Malaria Fever Therapy, 1910–1950," *Bulletin of the History of Medicine* 70 (1996): 577–608; Joel Braslow, *Mental Ills and Bodily Cures: Psychiatric Treatment in the First Half of the Twentieth Century* (Berkeley, CA, 1997).

6. Whitrow (n. 2 above). On the search for alternatives to malaria, see Albert Heyman, "The Treatment of Neurosyphilis by Continuous Infusion of Typhoid Vaccine," *Venereal Disease Information* 26 (1945): 2–8; and William Bierman, "The History of Fever Therapy in the Treatment of Disease," *Bulletin of the New York Academy of Medicine* 18 (1942): 65–75.

7. Jack R. Ewalt, Ernest H. Parsons, Stafford L. Warren, and Stafford L. Osborne, *Fever Therapy Technique* (New York, 1939), 3–4.

8. Paul de Kruif, *Men against Death* (New York, 1932), 249–79.

9. Theodore C. C. Fong, "A Study of the Mortality Rate and Complications following Therapeutic Malaria," *Southern Medical Journal* 30 (1937): 1084–88; William Kraus, "Analysis of Reports of 8,354 Cases of IMPF-Malaria," *Southern Medical Journal* 25 (1932): 537–41; S. P. James, "Epidemiological Results of a Laboratory Study of Malaria in England," *Transactions of the Royal Society of Tropical Medicine and Hygiene* 20 (1926): 148–65; S. P. James, "Some General Results of a Study of Induced Malaria in England," *Transactions of the Royal Society of Tropical Medicine and Hygiene* 24 (1931): 477–538.

10. James, "Some General Results" (n. 9 above).

11. Mark F. Boyd and Warren K. Stratman-Thomas, "A Controlled Technique for the Employment of Naturally Induced Malaria in the Therapy of Paresis," *American Journal of Hygiene* 17 (1933): 37–54.

12. Mark Boyd, Diaries, folders 468–470, box 48, Rockefeller Foundation Archives, RF 1, 100 Series, Rockefeller Archive Center, North Tarrytown, New York (hereafter cited as RAC). Archives director Darwin Stapleton and his staff made research in this archive consistently pleasant, efficient, and productive.

13. On the Tuskegee syphilis study, see James Jones, *Bad Blood: The Tuskegee Syphilis Experiment* (New York, 1981); and Allan M. Brandt, "Racism and Research: The Case of the Tuskegee Syphilis Study," *Hastings Center Report* 8 (1978): 21–29.

14. Boyd, Diaries, May 1931, RAC.

15. Ibid.

16. Boyd, Diaries, July 14, 1931, RAC.

17. Boyd, Diaries, July 7, 1933, RAC.

18. Boyd, Diaries, May and June 1931, RAC.

19. Boyd, Diaries, June and July 1931, RAC.

20. Boyd, Diaries, July 10, 1931, RAC.

21. Boyd, Diaries, July 26 and July 28, 1931, RAC.

22. Ibid.

23. Boyd and Stratman-Thomas (n. 11 above), 54.

24. Mark F. Boyd and Warren K. Stratman-Thomas, "Studies on Benign Tertian Malaria. 1. On the Occurrence of Acquired Tolerance to *Plasmodium vivax,*" *American Journal of Hygiene* 17 (1933): 55–59.

25. Mark F. Boyd and Warren K. Stratman-Thomas, "Studies of Benign Tertian Malaria. 2. The Clinical Characteristics of the Disease in Relation to the Dosage of Sporozoites," *American Journal of Hygiene* 17 (1933): 666–85.

26. Ibid., 683–84.

27. Philip Curtin, *Death by Migration: Europe's Encounter with the Tropical World in the Nineteenth Century* (Cambridge, 1989).

28. Frederick L. Hoffman, "Race Traits and Tendencies of the American Negro," *Publications of the American Economic Association* XI (1896): 1–139; Marvin L. Graves, "The Negro a Menace to the Health of the White Race," *Southern Medical Journal* 9 (1916): 407–13; C. Jeff Miller, "Special Medical Problems of the Colored Woman," *Southern Medical Journal* 25 (1932): 733–39; L. C. Allen, "The Negro Health Problem," *American Journal of Public Health* 5 (1915): 194–203.

29. Kenneth F. Maxcy, "Spleen Rate of School Boys in the Mississippi Delta," *Public Health Reports* 38 (1923): 2466–72.

30. For malaria statistics by race see, M. A. Barber, et al., "Prevalence of Malaria (1925) in Parts of Delta of Mississippi and Arkansas: Economic Conditions," *Southern Medical Journal* 19 (1926): 373–77; M. A. Barber and Bruce Mayne, "The Seasonal Incidence of Malaria Parasites in the Southern United States," *Southern Medical Journal* 17 (1924): 583–91; R. H. von Ezdorf, "Endemic Index of Malaria in the United States," *Public Health Reports* 31 (1916): 819–28; Mary Gover, "Negro Mortality," *Public Health Reports* 61 (1946): 259–65, 1529–38; ibid, 63 (1948): 201–13; ibid, 66 (1951): 295–305.

31. Martin D. Young et al., "Experimental Testing of the Immunity of Negroes to *Plasmodium vivax,*" *Journal of Parasitology* 41 (1955): 315–18; Louis H. Miller et al., "Erythrocyte Receptors for (*Plasmodium knowlesi*) Malaria: Duffy Blood Group Determinants," *Science* 189 (1975): 561–63; Louis H. Miller et al., "The Resistance Factor to *Plasmodium vivax* in Blacks: The Duffy-Blood-Group Genotype, FyFy," *New England Journal of Medicine* 295 (1976): 302–4.

32. D. J. Weatherall, "Common Genetic Disorders of the Red Cell and the 'Malaria Hypothesis,'" *Annals of Tropical Medicine and Parasitology* 81 (1987): 539–48; F. Fleming, "Abnormal Haemoglobins in the Sudan Savanna of Nigeria," *Annals of Tropical Medicine and Parasitology* 73 (1979): 161–72; L. Luca Cavalli-Sforza, Paolo Menozzi, and Alberto Piazza, *The History and Geography of Human Genes* (Princeton, 1994), 146ff.

33. A. C. Allison, "Protection Afforded by Sickle-Cell Trait against Subtertian Malarial Infection," *British Medical Journal* 1 (1954): 290–94; A. C. Allison, "The Distribution of the Sickle-Cell Trait in East Africa and Elsewhere, and Its Appar-

ent Relationship to the Incidence of Subtertian Malaria," *Transactions of the Royal Society of Tropical Medicine and Hygiene* 48 (1954): 312–18.

34. Boyd, Diaries, October 29, 1931, RAC.

35. Boyd, Diaries, November 21, 1931, RAC.

36. Mark F. Boyd and S. F. Kitchen, "Observations on Induced Falciparum Malaria," *American Journal of Tropical Medicine* 17 (1937): 213–35.

37. On the use of quartan malaria in neurosyphilis, see Bruce Mayne, "Note on Experimental Infection of Anopheles punctipennis with Quartan Malaria," *Public Health Reports* 47 (1932): 1771–73, and Bruce Mayne and Martin D. Young, "Antagonism between Species of Malaria Parasites in Induced Mixed Infections," *Public Health Reports* 53 (1938): 1289–91.

38. Mark F. Boyd, W. K. Stratman-Thomas, S. F. Kitchen, and W. H. Kupper, "A Review of the Results from the Employment of Malaria Therapy in the Treatment of Neurosyphilis in the Florida State Hospital," *American Journal of Psychiatry* 94 (1938): 1099–1114, quote at 1107.

39. Ibid., 1114.

40. On the treatment of nonhospitalized neurosyphilis patients, see Walter Freeman, "Therapeutic Malaria in Private Practice," *Southern Medical Journal* 24 (1931): 933–37.

41. Boyd, Diaries, January 13, 1934, RAC.

42. Boyd, Diaries, December 6, 1932, RAC.

43. On these trials, and the general history of malaria in the American South during the first half of the twentieth century, see my book, *Malaria: Poverty, Race, and Public Health in the United States* (Baltimore, 2001).

44. Boyd, Diaries, August 21, 1935, RAC.

45. Boyd, Diaries, November 20, 1932, RAC.

46. K. Mertz and K. C. Spitalny, "Imported Malaria Associated with Malariotherapy of Lyme Disease—New Jersey," *Morbidity and Mortality Weekly Report* 39 (1990): 873–75; "Self-induced Malaria Associated with Malariotherapy for Lyme-Disease—Texas," *Morbidity and Mortality Weekly Report* 40 (1991): 665–66. One physician in favor of studying the use of malariatherapy for central nervous system Lyme disease is Henry J. Heimlich. See his letter to the editor, *New England Journal of Medicine* 322 (1990): 1234–35. He also is involved in the Chinese AIDS effort. See Xiaoping Chen et al., "Phase-I Studies of Malariotherapy for HIV Infection," *Chinese Medical Sciences Journal* 14 (1999): 224–28; and Henry Heimlich et al., "Malariotherapy for HIV Patients," *Mechanisms of Aging and Development* 93 (1997): 79–85.

47. Wolfgang U. Eckart and Hana Vondra, "Malaria and World War II—German Malaria Experiments 1939–45," forthcoming in *Parassitologia*. The quotation is from Dr. Claus Karl Schilling, translated from the German by Eckart and Vondra.

48. William E. Collins and Geoffrey M. Jeffery, "A Retrospective Examination of Sporozoite- and Trophozoite-induced Infections with *Plasmodium falciparum*: Development of Parasitologic and Clinical Immunity during Primary Infection," *American Journal of Tropical Medicine* 61 (1999): 4–19; idem, "A Retrospective Examination of Secondary Sporozoite-and Trophozoite-induced Infections with *Plasmodium falciparum*: Development of Parasitologic and Clinical Immunity follow-

ing Secondary Infection," *American Journal of Tropical Medicine* 61 (1999): 20–35; idem, "A Retrospective Examination of Sporozoite-and Trophozoite-induced Infections with *Plasmodium falciparum* in Patients Previously Infected with Heterologous Species of Plasmodium: Effect on Development of Parasitologic and Clinical Immunity," *American Journal of Tropical Medicine* 61 (1999): 36–43; idem, "A Retrospective Examination of the Patterns of Recrudescence in Patients Infected with *Plasmodium falciparum,*" *American Journal of Tropical Medicine* 61 (1999): 44–48.

49. Charles Weijer, "Another Tuskegee?" *American Journal of Tropical Medicine* 61 (1999): 1–2, quote at 2.

50. Bruce Mayne and Martin D. Young, "The Technic of Induced Malaria as Used in the South Carolina State Hospital," *Venereal Disease Information* 22 (1941): 271–76, quote at 272.

51. See, for example, Marion Torchia, "Tuberculosis among American Negroes: Medical Research on a Racial Disease, 1830–1959," *Journal of the History of Medicine and Allied Sciences* 32 (1977): 252–79; A. G. Fort, "The Negro Health Problem in Rural Communities," *American Journal of Public Health* 5 (1915): 191–93; W. E. Burghardt DuBois, *The Health and Physique of the Negro American* (Atlanta, GA, 1906); James A. Doull, "Comparative Racial Immunity to Diseases," *Journal of Negro Education* 6 (1937): 429–37.

52. Braslow (n. 5 above).

Part II: Who Experiments?

Human Radiation Experiments and the Formation of Medical Physics at the University of California, San Francisco and Berkeley, 1937–1962

David S. Jones and Robert L. Martensen

In the fall of 1937 Gunda Lawrence lay dying of an abdominal tumor; her doctors at the Mayo Clinic "had given her up."[1] Her son John, however, had an idea. He had recently left the medical faculty of Yale University to join his brother Ernest, inventor of the cyclotron, in the Department of Physics at the University of California at Berkeley (UCB). There the brothers collaborated on medical applications of radioisotopes, a technology they believed would revolutionize medicine as much as had the microscope.[2]

On the basis of his recent cyclotron experiments involving irradiated tumors in mice, John Lawrence knew that tumor cells were especially susceptible to the toxic effects of radiation. He believed that radiation treatment might save his mother. Lawrence went to Minnesota and brought his mother, bleeding most of the way, by train back to California. Knowing that no conventional therapies could help her, and aware of the tremendous gain that would come from success, he aimed a million-volt x-ray tube, one of only two in the nation, into his mother's abdomen: "I had to practically stand over her and insist on more and more treatments in 1937. I would drive her over for each treatment when she became ambulatory and I would often have to stop the car on the way home so she could vomit. She pleaded with me to stop the treatments and at times I felt very cruel in not giving in—she however was cured—but it took ten years before I could get her to believe this."[3] Gunda Lawrence lived for twenty-two more years, during which she would see Ernest win a Nobel Prize in Physics for the cyclotron and John use his brother's invention to found a new field of medicine.[4]

Using archival material from UCB, the University of California Hospital and Medical School (now University of California, San Francisco, or UCSF), and the Lawrence Berkeley Radiation Laboratory (LBRL), this chapter explores the history of some human radiation experiments at the University of California. Four projects evolved in parallel: experiments on the biologic effects of radiation, studies of the effects of radiation on soldiers, attempts to develop radiation therapy for cancer, and use of radioactive tracers to study normal human metabolism. They show how a group of physicians struggled with the ethical implications of their work as they exploited a new technology to define a new field of medicine with themselves at the core. Our analysis emphasizes power and interest relationships of early medical physics researchers and sponsors at one institution. We want to situate their behavior, including their ethical transgressions, in the context of their resolute pursuit of the dual goals of establishing a new professional discipline and achieving preeminence in medical physics.[5] We argue that the professional and political interests of researchers and sponsors occasionally impinged on their beneficence.

We do not pretend to provide a definitive account of medical physics at the University of California or elsewhere; rather, we hope to stimulate discussion and additional historical analysis of the topic. For an extensive summary of U.S. human radiation experiments, readers should consult the *Final Report* of the President's Advisory Committee on Human Radiation Experiments (ACHRE) and its supplements.[6]

Experiments on the Harmful Biologic Effects of Radiation

When John Lawrence started working at his brother's laboratory at Berkeley, they had little concern about the hazards of radiation. Researchers casually exposed themselves to ionizing radiation, entering the cyclotron room to "look at this beautiful purple deuteron beam to get a glance of it. It was a beautiful thing; I saw it."[7] When he started doing his animal toxicity studies, they stopped watching. Over the 1930s and 1940s, as radiation science grew into an industry, physicians' uncertainty over the health effects of radiation became a major problem.

Scientists had known from the earliest studies of radioactivity that radiation could harm living tissues. Within a few years of Roentgen's 1895 discovery of x-rays, radiation had been shown to cause burns, ulcerations, and skin cancer. Studies of radiologists and radium dial painters showed the effects they suffered from their chronic exposure. However, the exact risks of radiation exposure eluded early scientists.[8] Two stories reflect their confusion. John Lawrence's first animal experiment at Berkeley exposed a rat to the direct deuteron beam of the cyclotron. After two minutes he turned off the cyclotron so he "could crawl back in there . . . and see whether the rat was okay, and the rat was dead! . . . I think quite honestly we didn't know for sure what killed the rat. It scared everybody."[9] Only after two weeks, when they had analyzed the pathologic sections, did they realize that the rat had actually died of suffocation—the air hose had become disconnected during the experiment. There was no evidence of any radiation damage. Meanwhile, Manhattan Project researchers raised mice in a project laboratory where uranium oxide dust was thick. They developed into healthier mice than controls in the animal farm.[10] Such mixed results, in which early animal studies did not confirm initial beliefs about the toxicity of radiation, left scientists confused and concerned.

In the late 1930s and early 1940s, this uncertainty became more significant as work on radiation expanded enormously and became organized along industrial lines. More and more researchers at universities began to work with cyclotrons, high-energy particles, and radioisotopes. The Manhattan Project, begun in 1942, exposed thousands of workers and scientists to the unknown effects of radiation. Project leaders repeatedly queried physicians concerning the biological effects of radiation in order to predict what doses would be safe, to learn how to protect people from dangerous doses, and to treat those who had been exposed. With so much depending on knowledge of the biological effects of radiation, scientists began extensive animal experimentation. When they found that different animal species excreted radioisotopes, such as plutonium, at different rates, they realized that animal results could not be reliably extrapolated to humans. They had to conduct human experiments.[11]

The complete scope of the human radiation experiments, from the

Manhattan Project through the cold war and into the 1970s, is still being established. The Department of Energy (DOE), for instance, has struggled to evaluate 3.2 million cubic feet of documents relating to the experiments.[12] Preliminary estimates show that more than 9,000 people were deliberately exposed to radiation in 154 experiments sponsored by the DOE alone.[13] The most infamous experiments are the plutonium injection studies conducted from 1945 to 1947 at the universities of Rochester, Chicago, and UCSF, in which eighteen patients were injected with plutonium without their knowledge.[14] In similar studies, researchers injected uranium into patients to measure the dose at which detectable kidney damage began.[15] Other researchers, at UCSF and Berkeley, exposed subjects' skin to radioactive phosphorus to correlate dose-hours with burn damage.[16] In Oregon, they irradiated the testicles of "volunteers" to measure the effect of exposure to x-rays on sperm production.[17] Elsewhere, they estimated the health consequences of a nuclear war by feeding radioactive fallout from the military's Nevada test site to healthy volunteers.[18]

What sorts of people would serve as subjects for these studies? Some subjects, especially in the most toxic experiments, had no choice. The thirteen patients used in the two uranium studies all suffered from terminal cancer; ten of them were comatose or semi-comatose at the time the studies were initiated.[19] Other subjects came from captive populations. The "volunteers" for the testicular irradiation were inmates at the Oregon State Prison who agreed to participate in the study and were paid more than one hundred dollars for their assistance.[20] The people who agreed to ingest radioactive fallout were students and staff at the University of Chicago "who hoped that they were making a contribution to Civil Defense."[21]

It is difficult to tell today the degree to which the studies harmed any subjects. Historically, however, the question of actual harm is less interesting than the researchers' belief about the risk to which they exposed their subjects. Both their choice of research populations and the precautions they took reflect the expectation that the subjects would be harmed. In some cases, such as the uranium injection studies, the explicit purpose was to measure the harm caused.[22] Researchers minimized the consequences by using patients whom they believed to be terminally ill.[23] This was explicit in the plutonium studies: "I feel reasonably certain that there would be no harm in using larger amounts of [plutonium] if you are sure the case is a terminal one."[24] Unfortunately, these physicians had limited prognostic powers: four of the eighteen "terminally ill" recipients of plutonium lived for more than twenty years.[25] Furthermore, they often sought dying patients who were "relatively healthy," with intact kidney and liver function: it was "desirable to obtain a metabolic picture comparable to that of an active worker."[26] In other cases, as in the testicular irradiation studies, researchers took precautions to prevent long-term consequences from damage that irradiation might cause. They required that every prisoner who participated undergo a vasectomy following the experiment.[27]

The researchers' efforts to minimize the damage from radiation confirm their belief that their experiments could harm the subjects. Previous commentators on cold war research have tended to excuse the ethical transgressions. In his classic 1966 study on unethical research, Henry Beecher wrote that "thoughtlessness and carelessness, not a willful disregard of the patient's rights, account for most of the cases uncovered."[28] The Markey Report, a 1986 congressional investigation of some of the radiation studies, naively dismissed this issue by stating that: (1) the experiments occurred during the less "enlightened" 1940s and 1950s; (2) they "might be attributed to an ignorance of the long-term effects of radiation exposure"; and (3) "the sad history of human radiation experimentation makes it clear that standards that were acceptable forty years ago appear repugnant today."[29] A 1991 assessment of "scientific data obtained by immoral means" was similarly forgiving. While such research might be "unethical if done now," the "prisoners had the chance to serve humanity and the soldiers were helping to learn how to win wars." All subjects knew the risks: healthy subjects had volunteered, while sick subjects participated in therapeutic studies. As a result, the studies were "ethical if the design was valid and the results likely to be helpful to mankind."[30]

However, the documentary evidence does not support such ahistorical responses. The radiation experiments did not occur in a historical vacuum, nor did discussions of the ethics of human experimentation only begin after the experiments had finished. In *Subjected to Science* (1995) Susan Lederer shows that harmful human experimentation and sophisticated discussions of the ethical implications of such work have existed since the nineteenth century.[31] Researchers had exposed orphans and patients at charity hospitals to pathogenic microbes. Walter Reed exposed American soldiers and Cuban volunteers to mosquitoes carrying yellow fever (3–26). By the turn of the century many American doctors opposed any research that was not explicitly for the benefit of the patient. In 1907, for instance, William Osler argued that physicians should only experiment on patients once the "absolute safety" of the procedure had been demonstrated by animal studies, only with the "full consent" of the patient, and only if "direct benefit of the individual is likely to follow" (1). In 1916, Walter Cannon and other prominent researchers tried to amend the American Medical Association's Code of Ethics to include a statement requiring patient consent in research (97–100).

In the 1940s, therefore, the consensus of the medical profession, albeit unlegislated, was that only experimental studies expected to be therapeutically beneficial should be done on patients. Moreover, by 1951 a series of ethical guidelines had been promulgated for medical research. In 1946, both the War Crimes Tribunal at Nuremberg and the American Medical Association required the consent of research subjects.[32] In 1947 and again in 1951, the Atomic Energy Commission required that all radiation experiments have the "hope of therapeutic benefit."[33] Despite these guidelines, researchers continued with many human radiation experiments, known to be toxic, without expected benefit, violating both written and unwritten ethical principles of medical practice.

The scientists involved in the experiments knew that they were ethically in shadowy terrain. Far from the unenlightened bumblers supposed by the Markey Report, these scientists knew of their transgressions and responded in the safest way they knew: they classified their experiments, effectively blocking them from the public's eye for nearly fifty years. In 1947 a memo from the Manhattan Project to the Atomic Energy Commission (AEC) stated: "It is desired that no document be released which refers to [the plutonium injection experiments] with humans and might have adverse effect on public opinion or result in legal suits. Documents covering such work should be classified secret."[34] Another AEC official wrote later that year that there were "a large number of papers which do not violate security but do cause considerable concern to the Atomic Energy Commission insurance branch, and may well compromise the public prestige and best interests of the commission."[35] Concerned about possible public outcry and lawsuits that would occur if the public ever learned of the unethical radiation experiments, the researchers involved declared them "Top Secret" and left the evidence to molder in remote federal archives.[36]

Studies of the Effects of Radiation on Soldiers

The destruction of Hiroshima and Nagasaki by atomic bomb in August 1945 provided the first, and most dramatic, public demonstration of the power and the risks of radiation. As World War II moved into the cold war, the military continued to dominate radiation research through its funding of experiments on the effects of radiation on soldiers. The Berkeley Radiation Laboratory again played a major role. Berkeley scientists had been involved with the military radiation project from its outset. Using the cyclotron, they had produced the first samples of uranium-235, the isotope used in the Hiroshima bomb. They had discovered plutonium, the element used in the Nagasaki bomb.[37] They had used radioisotopes to study the physiology of pilots' exposure to high altitude and divers' exposure to high pressure and decompression.[38] These projects served as a significant source of income for the laboratories. When the Division of Medical Physics was established in 1945, the University of California provided an annual budget of $11,350.[39] Meanwhile, the Army paid $15,370 for the decompression study alone.[40] The military also paid $3,900 of Lawrence's $4,500 annual salary.[41]

Most of the military research uncovered at Berkeley was led by Joseph Hamilton. Trained in chemistry as an undergraduate at Berkeley and in medicine at UCSF, Hamilton became a physics fellow at the Crocker laboratory.[42] On Christmas Eve 1936, he attempted the first "therapeutic" use of radioisotopes, treating leukemia with radioactive sodium.[43] He continued this work and, from 1942 until 1948, directed the radioisotope metabolism studies for the Manhattan Project. He personally oversaw the plutonium injection studies performed at UCSF, including the last one, performed in 1947, after he had been ordered by the Manhattan Project to stop.[44] He continued his work on the metabolism of radioisotopes until his untimely death in 1959, at the age of 49, from a rare form of leukemia.[45]

Using his expertise on the biological effects of radiation, Hamilton produced a remarkable report for the military on the potential "military applications of fission products."[46] He began by noting the special toxic properties of radioactive materials that made them useful weapons: microgram doses produced lethal effects, smaller doses produced incurable tissue damage with a long latency and chronic effects, special devices were needed to detect the presence of radiation, and their effects on land and buildings persisted for a very long time. Hamilton calculated inhaled and ingested toxic doses and proposed devices that could be used to disperse the materials over large areas. He even described how the weapons might be used: "One of the principal strategic uses of fission products will probably be against the civilian population of large cities. It can be well imagined the degree of consternation, as well as fear and apprehension, that such an agent would produce upon a large urban population after its initial use" (6). Tactically, they could be used to contaminate specific areas, blocking access to beachheads and mountain passes. Furthermore: "The direct use of fission products either against massed troops or against personnel in trench fortifications, not readily neutralized by more conventional agents, might be quite effective . . . [and] the use of fission products might be effectively extended against naval vessels either by shelling, using missiles containing fission products, or by spraying the material over the entire ship" (6).

After cataloging such a frightening array of military applications of radioisotopes, Hamilton concluded that "the best protection that this nation can secure against the possibilities of radioactive agents being employed as a military tool by some foreign power is a thorough evaluation and understanding of the full potentiality of such an agent" (7). This report shows that physicians conducting radiation experiments did not confine themselves to collecting data on biological effects. At times they also became proactive and used their data derived from experiments with laboratory animals, comatose patients, and "volunteers" to provide military leaders with expert advice on ways to use radioactive poisons most effectively. However, Hamilton believed that the actual work of weapon development ought to be done by the military, with civilian oversight. He believed it "inappropriate" for universities to be involved in such research (7).

This distinction between military and university-civilian spheres of knowledge is mirrored in another of Hamilton's interactions with the military: his role as a consultant to the Army's experiments on the effects of atomic blasts on the functioning of soldiers. As the Korean War progressed and the tactical use of atomic bombs became a real possibility, the military developed new concerns about atomic weapons. While civilian scientists studied what radiation exposure could be tolerated by workers without harm, military leaders explored what exposure could be tolerated by soldiers without compromising performance. Suppose two armies faced each other several miles apart. Army commanders needed to know what would happen to their own soldiers if a tactical atomic bomb were dropped on the enemy ranks. The closer they could station their soldiers to ground zero, the better they could take advantage of the ensuing confusion among enemy soldiers. But how close could soldiers be to the vicinity of a blast without losing their ability to function?

To answer this question, the military exposed personnel to atomic blasts in Nevada. Some studies calibrated radiation detection devices, developed protective gear, or tested decontamination procedures. Others observed physiological and psychological effects of being near ground zero.[47] One study examined "the effects of the flash of atomic detonations at night upon the ability of military personnel to carry out assigned tasks involving the use of vision." Volunteers, some wearing red goggles, others with eyes unprotected, watched a blast and then waited to see how long it took—as long as eleven minutes—until they could read their instruments. A few volunteers developed retinal lesions, which subsequently healed.[48]

In another case, a group of carefully selected volunteer officers were stationed exceptionally close—only 2000 yards—to ground zero. Before participating, each officer had to "personally and individually compute the effects to be expected" from witnessing a 40–kiloton blast from that proximity. One officer produced an astonishing eyewitness account of the event, including descriptions of the white flash, shock wave, and ensuing radioactive fallout. He describes that while the trenches protected the soldiers from any immediate damage, nearby Joshua trees had burst into flame and sheep tethered nearby "were singed to a dark brown color" but were still standing without other signs of injury.[49]

The military was not ignorant of the ethical issues surrounding human experimentation. It conducted the studies on the recommendation of the Armed Forces Medical Policy Council: "Human subjects [should be] employed, under recognized safeguards, as the only feasible means for realistic evaluation and/or development of effective preventive measures of defense against atomic, biological, or chemical agents." A 1953 memo to the Secretaries of the Army, Navy, and Air Force provided guidelines for such studies. All human experiments were to be based on prior animal experiments, use strict standards of written informed consent, and minimize the "physical and mental suffering and injury."[50]

Military leaders sought the advice from project physicians about what levels of exposure would be safe. In some cases, it was advice the military did not wish to hear. For example, in 1952 Dr. Shields Warren, director of the Division of Biology and Medicine of the Atomic Energy Commission, advised the military not to station troops within seven miles of ground zero. He was concerned because the yield of experimental explosions "cannot be predicted with accuracy." In addition, "any injury or death during the operation might well have serious adverse effects" on public relations.[51]

The military ignored this advice and positioned officers within 2000 yards (just over one mile). Joseph Hamilton had raised similar concerns in 1949 about testing radiological weapons. His objections had also been ignored, and the tests, which released substantial radiation into the atmosphere, had proceeded as planned.

In Hamilton's work as a consultant to military weapons designers, he had established a boundary between "appropriate" settings for research. On one side was research that could be done by universities; on the other was work that the military would have to conduct with its own resources. In his work as a health consultant, he, like Shields Warren, tried to define boundaries between what levels of radiation exposure were permissible for research subjects. On one side were experiments that were permissible, and on the other were experiments that should not be permissible, even for the military.

The military, however, could (and did) ignore this kind of advice. A striking debate between Warren and military leaders shows how the different professional roles of doctors and generals produced their divergent views on the role of human experimentation. General James Cooney, head of the radiological branch of the AEC's Division of Military Application and a forceful advocate of performing atmospheric tests in the continental United States, noted that military leaders needed to know the effect of radiation on soldiers' performance: "[At] every conference I attend of the military, I am asked by the line officer how much radiation can a man take?"[52] Admiral Greaves argued that a few experiments, which might harm hundreds of soldiers, were preferable to waiting until combat, when weapons would be tested on thousands of troops: "[It] is going to be more economical in the long run to take a few chances now and perhaps not lose a battle or even worse than that, and not lose a war" (17). Cooney saw matters simply: "Personally, I see no difference in subjecting men to this than I do to any other type of experimentation that has even been carried on. Walter Reed killed some people. It was certainly the end result that was very wonderful. Shall we wait until we find out and force people and force thousands of young men perhaps and maybe lose the battle as a result of not knowing, and so on?" (7–8).

For the military, the ends justified the means. But many of the physicians involved had a different perspective on human experimentation. Shields Warren, echoing William Osler's statements from 1907, believed that exposing soldiers was not justified because the potential of animal experiments had not been exhausted. He was "very much opposed to human experimentation when it isn't for the good of the individual concerned and when there is any other way of solving the problem" (15). Moreover, he argued that even a study of hundreds of soldiers would not provide statistically significant data (13). The effects of radiation varied from person to person and appeared to depend on the soldiers' level of rest or activity and amount of clothing and protection from the blast. Nothing short of massive experimentation could provide the generals with the certainty they wanted. For Warren and his colleagues, not only did the ends not justify the means, but the means would not even produce the desired ends. The experiments were both unethical and pointless. In the end, the military took its own counsel, conducted the studies, and classified all papers related them, including the doctors' objections, as "secret."

Attempts to Develop Radiation Therapy for Cancer

Before UC Berkeley investigators became involved with radiation-based weapons, they concentrated on the potential of radiation to kill tumor cells. Soon after John Lawrence arrived at Berkeley, he started a series of experiments in which he took tumors out of mice and then irradiated and reimplanted them. In the absence of radiation, the replanted tumor would be 100 percent lethal. However, radiation damaged the tumors; and at increasing doses, an increasing percentage of mice survived. John and Ernest found that neutrons were about 20 percent more lethal to tumor cells than to normal cells: "We and others said that maybe we had a radiation here that is selective against cancer."[53] By 1938 they began treating patients with the cyclotron's neutron beam.[54] Researchers at Berkeley and UCSF and many other institutions quickly began comprehensive programs to develop radiation therapy for cancer. This work raised now familiar ethical questions: What sorts of people could be used as research subjects? What levels of experimental risk were acceptable? Just as the military argued that the cold war justified more dangerous experiments, some cancer researchers argued that the desperate problem of cancer justified more flexible ethics.

The cancer research had two basic components. First, scientists used animal and human experiments to describe the biologic effects and measure the toxicity of radioisotopes. Some of these experiments have been described above. The uranium injection study, for instance, was done to develop potential therapies for cancer: "The determination of an intravenous tolerance dose is an initial step in evaluating the possible application of U_{235} to the neutron capture therapy of brain tumors."[55] Second, once the biological effects had been described, researchers proceeded to therapeutic trials. Bertram Low-Beer, the UCSF radiologist who conducted the phosphorus studies, began therapeutic trials as early as 1941. He treated hundreds of lesions, ranging from common warts to basal cell carcinoma, with P_{32} obtained from the Berkeley cyclotron. Success rates ranged from 88 percent for warts to 98 percent for basal cell carcinomas and 100 percent for hyperkeratosis.[56]

Such early successes contributed to the organization of what for a time was the country's largest cancer research facility at UCSF. During the 1940s the University of California had expanded its cancer research efforts and sought grants from the state, from the federal government, and from private donors. At the same time, Congress increased funding for cancer research and education. Since the National Institutes of Health (NIH) campus at Bethesda, Maryland, was not yet able to accommodate the demands of cancer research, the NIH established regional cancer centers. These efforts converged to establish the Cancer Research Institute (CRI) at UCSF in 1947. This program had remarkable funding: between private, state, and federal donors, the CRI had a budget of $620,000 in 1947 with a $1 million NIH grant for new building.[57] In comparison, in 1947 the entire NIH budget for extramural grants was only $4,004,000.[58]

The CRI oversaw a broad program that explored all aspects of cancer, from pathology and cell biology to diagnosis and therapy. To direct research on new therapeutic agents, the CRI founded the Laboratory of Experimental Oncology (LEO), with a 1947 budget of $212,000, under the direction of Michael Shimkin, a professor of oncology at UCSF and an officer in the Public Health Service. Working primarily at the Laguna Honda Home "for the aged poor," Shimkin tested many chemotherapeutic agents, ranging from nitrogen mustards to hormones and digestive enzymes.[59]

Two particular studies received condemnation from Shimkin's colleagues. The first involved homologous melanoma transplants, in patients with metastatic disease, to study the number of cells required to establish a metastasis at a distant site.[60] The second involved radiation. Some researchers had proposed that leukemic cells came from the bone marrow. Shimkin wanted to test this by using massive full-body radiation to destroy the marrow of leukemic patients. However, he knew that bone marrow ablation would be fatal, so he planned to keep the patients alive by co-transfusing them with other leukemic patients.[61]

All of the research proposals of the LEO had to be approved by the UCSF Cancer Board. These two proposals were not well received. Frequently citing the guidelines from the War Crimes Tribunal at Nuremberg and the AMA's Code of Ethics, members of the Cancer Board questioned whether the experiments were "in line with the basic principles governing human experimentation." Dr. Robert S. Stone, chair of the Cancer Board and chair of radiology at UCSF, opposed the melanoma study: "Even a patient with untreatable cancer should not be subjected to another malignancy." Shimkin defended his plans: although full body irradiation was "a hazardous procedure and mortality is expected," he said, the "aim is therapeutic." Stone disagreed. As long as he controlled access to his department, he declared, "This will not be done with x-ray at the University of California." The Board denied permission for both experiments.[62]

This rejection of the LEO's research program was not an isolated event. By 1951, Shimkin had been accused by the director of NIH of "vivisecting on man."[63] In 1953, as the 500–bed NIH clinical center opened in Bethesda, Shimkin received notification from UCSF and the CRI that the clinical research at the LEO would no longer be supported. They wrote that the decision was "based on budgetary considerations and should not be interpreted as a reflection on the quality of the work."[64]

Like the military, who believed that cold war exigencies justified unethical experiments, Shimkin believed that his patients' hopeless condition allowed different standards of ethical conduct. However, unlike the military, which had the power to proceed with the studies despite Warren's and Hamilton's objections, Shimkin could not continue without the support of the Cancer Board. The board had established boundaries that defined which experiments were ethical and which were not. This was a crucial step in the self-definition of the new specialty of medical physics.

Radioisotope Tracer Studies of "Normal" Human Metabolism

Thus far this chapter has focused on the eye-catching experiments in which researchers rationalized their rejection of accepted codes of ethical research. The histories of the experiments on the biologic effects of radiation, the studies of the effects of radiation on soldiers, and the attempts to develop radiation therapy for cancer all reveal that scientists knew they were working in the ethical shadows. The radioisotope tracer studies of human metabolism did not raise such obvious ethical concerns. However, they again challenged the researchers to establish standards for what types of research would be acceptable and on what kinds of subjects.

As described earlier, the cyclotron allowed researchers to convert elements into their radioactive isotopes, which then could be used in minute doses as tracers. Radioactive sodium could be followed from its absorption by the intestine to its distribution throughout the body. Different labeled elements, such as potassium or chlorine, highlighted different metabolic processes. This made invisible processes visible and gave researchers a new window into physiology and body composition. Other groups shared the excitement with this new type of research. When the Manhattan Project announced its isotope distribution program in 1946, such tracer studies, "publishable researches in the fundamental sciences, including human tracer applications, requiring relatively small samples," received top priority.[65]

Early work, usually on animals, developed the new techniques. Joseph Hamilton, for instance, had generated enough data from rats by 1949 to publish a review article in the *New England Journal of Medicine*.[66] But as happened with cancer studies, researchers quickly began experimenting on humans. In one early study, Hamilton had eight "normal" subjects drink solutions containing radioactive sodium, potassium, chlorine, bromine, or iodine; he then used a Geiger counter to measure the amount of radiation they absorbed. He hoped to establish average values in healthy patients to compare to experiments in patients with abnormal physiology.[67] Other work at Berkeley searched for isotopes that concentrated in specific tissues, as iodine does in the thyroid. The choice of research subjects for such studies, in contrast to the toxic experiments described earlier, suggests that the researchers believed that their research would be harmless. One study used medical students who granted informed consent.[68] Another used ambulatory patients, aged 7 to 76, from the researchers' clinical practice.[69] A third study used workers from Los Alamos and even the wife of the principal investigator.[70]

As Hamilton had described, this metabolic research sought to establish normal values in health and deviations in disease. The researchers, however, had a difficult time finding enough "normal" people (i.e., healthy adult men) for their studies. One source was the military: researchers at

Berkeley collaborated with the U.S. Naval Radiological Defense laboratory to measure normal values for body water and fat content in sailors.[71] They also studied "400 healthy male railway workers."[72] There was even discussion, and possible use, of professional football players from the San Francisco 49ers.[73] Despite such resourcefulness, Berkeley scientists still needed more healthy volunteers: "We have now done a considerable number of these determinations on abnormal groups of patients, but have not had a normal group of subjects with which to compare our results."[74]

The researchers turned to prisons. Lawrence wrote that they needed to use inmates because they were "unable to procure adequate numbers of normal individuals who are willing or able to take a complete day away from their occupation while the study is carried out."[75] Prisoners provided a captive population of healthy, mostly white males, who were guaranteed to be available for follow-up studies.[76] Starting in 1949, Lawrence arranged a collaboration between Berkeley and San Quentin Prison. He assured the prison medical officials that the studies would be harmless: the radioactive materials had been safely used "on a large number of patients in addition to several M.D.s" at Berkeley.[77] Although the comptroller of the University of California at Berkeley was concerned about the "possible legal complications" of this work, the research began by 1950.[78] One study used labeled chromium to measure both the iron content of blood and the blood volume of the inmates.[79] Another used tracers to measure total body water. A third studied the effects of the recently discovered hormone erythropoiten.[80] Such research involved only the injection of the material and the collection of blood samples for analysis. Other schemes used inmate volunteers as bone marrow donors for the treatment of leukemic children.[81] The donors were paid $25 to $50 for their participation; meanwhile, the prison guards who transported them received a "salary" of $25 to $35 plus expenses per trip.[82]

While the scientists as Berkeley used prison inmates to establish "normal" ranges for their metabolism studies, they also studied various "non-normal" conditions. Some work measured differences in total body water and fat content of individuals ranging from thin to "the occasional obese friends that you could pick up."[83] Other work involved patients with various diseases, including cardiogenic edema, diabetes, anemia, polycythemia, and leukemia.[84] Meanwhile, researchers at other universities, in consultation with the Berkeley scientists, planned or conducted work on pregnant women, healthy infants, and premature babies weighing as little as two pounds.[85] Some work used stillborns or aborted fetuses to study the distribution of isotopes in the fetus.[86]

Radioactive isotopes provided scientists with a new tool for exploring the human body, for studying processes that previous researchers could only guess at. Just as the microscope had generated a whole new picture of human anatomy and pathology, radioisotopes provided unprecedented information about human physiology and pathophysiology. This capacity became a tremendous temptation for the researchers. They wanted to study normal physiology and how it changes across the normal range of human variation. They also wanted to study disease states, or newborns and fetuses. Just as toxic effects of radiation prompted military and cancer research, harmless effects of radiation prompted tracer research. Where the former violated the then-accepted ethical standards by performing harmful experiments, the latter violated them by using vulnerable research populations.

Building the Empire

The history of medical uses of radioactive substances may be divided into pre- and post-cyclotron periods. Although the Hungarian chemist Georg von Hevesy in the 1910s and 1920s pioneered the use of radioisotopes as tracers in plant physiology, his materials, like those of all radiation researchers prior to the cyclotron, were limited to naturally occurring radioisotopes, all of which were (and are) rare and expensive.[87] As a result, when John Lawrence went to Berkeley in 1937, the field of radiation medicine and medical physics involved little more than a few experiments he planned to do with his brother's cyclotron. Within twenty-five years, everything had changed: radiation medicine had become an established specialty, and programs in medical physics existed worldwide. The Berkeley team did not miss this opportunity. Quick to recognize and exploit the potential of the cyclotron, they pursued an entrepreneurial campaign of academic expansion to establish and assert their dominance in the field.

The cyclotron, a virtual factory for creating isotopes out of natural elements, provided the key to the radiation medicine. Lawrence could pick an element he wanted to study, bombard it with neutrons, and produce a radioactive tracer for physiological studies. This access let the researchers explore new ground, placing them on the cutting edge of a new science. When John Lawrence, while still at Yale, had told Harvey Cushing about the opportunity at Berkeley, Cushing encouraged him to go. He "saw the excitement of this field. . . . He said that this was something like bacteriology when he was a young fellow; he said this was going to be a terrific field."[88] The Lawrences had been told that isotopes would "be as important as the microscope to the future of biology and medicine."[89] John felt the hope and excitement at the Berkeley Radiation Laboratory as soon as he arrived: "Ernest and I talked about the tremendous potential. . . . We talked about it a lot. We never figured we could do it. . . . It was really very exciting for me to come to a place where you were surrounded by all these basic scientists; all these bright young guys who could help you do things that you couldn't do."[90]

This excitement was compounded by the knowledge that they could do things that no one else could. The cyclotron (and eventually its technological successors such as the linear accelerator) held the key to both high-energy physics and radiation medicine. Until 1946, when the newly formed AEC began its radioisotope distribution program, only groups with access to a cyclotron and its dedicated team of physicists could be players in the field of radiation medicine. This limited research to Berkeley in the early 1930s, and then MIT and Harvard in the later 1930s. The era of "big science" had arrived, in medicine as well as in physics. Such work required substantial funding, and the Lawrence brothers (as well as Niels Bohr in Europe) were adept in using medical-biological arguments to gain governmental funding for cyclotron research.[91] Their early work on cancer was their selling point. Echoing the statements of early-twentieth-century radium therapists, they argued that post-cyclotron radiation medicine would provide a cure for cancer. As one co-worker would later describe, they "started the whole field, I guess, of the use of nuclear particles, and the particles in the treatment of cancer."[92]

Two obstacles, however, stood between the Berkeley Radiation Laboratory and its hopes to develop the full potential of radiation medicine: the faculty of the UCSF medical school wanted to restrict the laboratory's access to patients, and the Atomic Energy Commission (AEC) wanted to regulate the nature of the laboratory's experiments. To circumvent UCSF, Lawrence fought a campaign to establish a department of medical physics, complete with its own hospital unit, at Berkeley. At the same time, he had Hamilton use the laboratory's status as the founder of the field to convince the AEC to grant the laboratory special privileges.

Lawrence wanted a unified research space with intellectual independence. Knowing that he could never build a cyclotron in San Francisco, he needed to move the patients to Berkeley. First, he transferred his appointment, and his salary, from the medical school to the new Division of Medical Physics of the Department of Physics.[93] Meanwhile, as the success of his early experiments led into cancer research and military work, the division received expansive financial support from both the government and private sources, such as the Donner family.[94] These developments gave Lawrence the leverage he needed to fight for his own fiefdom. Arguing in 1947 that direct access to hospital patients would facilitate research and collaboration with the basic sciences at Berkeley, Lawrence petitioned to open the Donner Pavilion, a radioisotope clinical research unit.[95]

The UCSF faculty adamantly opposed Lawrence's proposal. The faculty worried about the risks of shuttling patients between San Francisco and Berkeley. They argued that "the medical staff [at Berkeley] is not adequately trained and that in certain instances human studies are conducted by non-medical members of the staff at Donner Laboratory."[96] They also feared loss of institutional integrity. The chair of the Department of Medicine said he was "considerably alarmed about the prospects of the 'Medical School going piece meal to Berkeley.'" It would create a schism between UCSF and Berkeley, producing the "appalling condition of having the Medical School divorced from its related sciences." The medical school proposed that Lawrence, like Shimkin and his LEO, use hospital beds in San Francisco.[97]

Lawrence fought such objections for years. He argued that other medical schools, notably Harvard, thrived despite being dispersed at multiple hospitals. He secured guarantees of financial support from Shields Warren at the AEC.[98] He argued that the medical school was meddling where it did not belong: "The problem concerned was not the problem of the Medical School, but the problem of research of the Division of Medical Physics."[99] Lawrence's efforts eventually paid off. The Donner Pavilion opened in 1952. A high-ranking committee subsequently agreed with Lawrence that UCSF should have no involvement in the clinical research at Berkeley. It recommended that "both the Division of Medical Physics and the Donner Laboratory, with its status as an Institute, be completely separated from the Department of Medicine of the School of Medicine."[100] Lawrence had his independent clinical research program at Berkeley.[101]

In the meantime, Lawrence had Hamilton work to protect Berkeley from federal regulation. In 1946, to utilize the "peacetime potentialities" of the Manhattan Project, the AEC began to distribute isotopes for research.[102] They also began to regulate the research: the AEC established its own review board and required that every institution that used isotopes establish an internal review board. This was unwelcome interference for the researchers at Berkeley, who had been using isotopes on their own for a decade. Hamilton pleaded with the AEC for special treatment. He believed that Berkeley's long experience with isotopes, predating even the Manhattan Project, placed it in a "considerably different position from almost any other large institution in the country." Because of their experience, he argued, "it is unlikely that members of these departments will request substances from your organization which are ill-considered, and under circumstances which might lead to potentially serious health hazards and problems." Even an internal review board would be offensive: "Not a few of our more experienced and senior faculty members using radioactive materials would somewhat resent the formation of a committee within the institution that would act as a board of review for their research problems."[103]

The AEC did not accept Hamilton's initial arguments. In 1949, Shields Warren issued a special statement to laboratories, including Berkeley, which could produce their own isotopes. He reiterated that the AEC had authority over all isotope experiments in humans, regardless of whether the isotopes had been shipped from the AEC or produced locally: "Allocations would be made by the Isotopes Division [of the AEC] only after review and approval by the Subcommittee on Human Applications of the [AEC's] Committee on Isotope Distribution. It should be emphasized that the instruction applies even though the radio material is produced in the laboratory where it is used. Since this procedure has not been uniformly followed in the past, we are writing to acquaint you with the appropriate details."[104]

Hamilton continued to fight. In a concession, Berkeley established an internal review board that made "full use of local expert talent," including physicians, chemists, physicists, and Lawrence, "the recognized dean of internal radiation therapy and a pioneer in the field of biological and medical uses of radioactive isotopes." Implying that their committee was as experienced as any could be, he argued AEC oversight was unnecessary. They were "certain that in the best interest of safe use of isotopes, that this committee of the Radiation Laboratory can assume the function of approval (or disapproval) of use."[105] The AEC acquiesced. Berkeley was "given permission by the Atomic Energy Commission to pass on all local use of isotopes in human investigations."[106]

By the early 1950s, the Berkeley Radiation Laboratory had established its independence from both federal control and the medical school at UCSF. As it established sovereignty, it also sought hegemony. First, it collaborated with researchers across the United States (NIH, Eastman Kodak, General Mills, United Air Lines, the Framingham Heart Study, San Quentin Prison) and overseas (London, Canada, Sweden, Brazil, Bombay, and Havana).[107] Berkeley-based researchers traveled to these sites to complete projects, provide advice, inspect facilities, or establish new programs.[108] Second, it trained a legion of researchers who established medical physics programs, modeled on Berkeley, throughout the world. By 1953 Berkeley trainees directed radiation medicine laboratories in Los Alamos, Walter Reed Hospital, and the Mayo Clinic, among other institutions, from San Francisco to San Diego, Seattle, New York, London, and Paris. Lawrence could confidently boast: "I don't think there is any place in the United States that has trained as many men in the clinical use of isotopes as we have trained here. . . . [A]s far as the safe use of radioactive isotopes in clinical investigation is concerned, I don't suppose there is a more experienced place in the country than right here in the Donner Laboratory."[109]

These radiation missionaries allowed the Berkeley researchers to influence and direct radiation research on a global scale. By training the people who then established radiation medicine throughout the country and the world, the Berkeley Radiation Laboratory established itself at the center of the new field. If, in the early days of his research at Berkeley, Lawrence had felt a sense of manifest destiny, then he succeeded in following it through to hegemony.

Conclusion

Between 1937 and the early 1960s, researchers at Berkeley used a new technology—radioisotopes generated by cyclotrons—to establish a new field of medicine. Their program had its growing pains. They and their colleagues at UCSF had to define standards of professional conduct, as seen in the ethical debates surrounding Shimkin's cancer research. They had to create a distinct specialty, as seen in their carving a niche at Berkeley separate from the department of internal medicine at UCSF. They had to use their stature, based on their long experience, to preserve their autonomy from the AEC. They also used their stature to attract graduate students, whom they then sent out to colonize hospitals and research laboratories throughout the world. Through these efforts, the Berkeley researchers established themselves as leaders of the new field of medical physics. All of their human subjects research took place on this stage of institutional ambition.

Most extant histories of medical physics, notably ACHRE's massive *Final Report*, emphasize past facts: the who, what, where, and when of experiments, trials, patient consent forms, waivers, government directives, and so forth. In doing so, they provide an invaluable resource for those seeking to trace radiation medicine's past. Our work has sought to deepen this understanding by exploring power and interest relationships of early medical physics researchers at one institution. This analysis can be generalized. The failure of biomedical researchers and their sponsors to acknowledge adequately the importance of personal, professional, and institutional ambition in shaping research behavior remains a persistent threat to the respect for patients and their interests in human subjects research in the United States. The pretense of ingenuousness, which forms a considerable portion of the public face of biomedical science, obscures some dimensions of what is at stake in past and present discussions about the aims and conduct of publicly funded biomedical research.

NOTES

1. John H. Lawrence (hereafter JHL), Letter to Herbert Childs, 13 July 1966 (Lawrence Berkeley Laboratory Archives and Records Office [hereafter LBLARO]: Scientists' Papers: John Lawrence Papers [hereafter JLP]: accession #434-90-AO186, file #19-146, carton #12, folder "Childs, Herbert").

2. JHL, "Isotopes and Nuclear Radiations in Medicine, or A Quarter Century of Nuclear Medicine," lecture delivered at the International Conference on the Use of Radioisotopes in Animal Biology and the Medical Sciences, sponsored by the International Atomic Energy Commission, the Food and Agriculture Organization of the UN, and the WHO, Mexico City, 24 November 1961 (JLP: #434-92-0277 / #19-14-6 / #3 of 6 / "speeches").

3. JHL, Letter to Childs, 13 July 1966.

4. "Questions for JHL," undated (JLP: #434-92-0066 / #19-14-6 / #2 / "N").

5. For a more general discussion of this tension, see David Rothman, "Other

People's Bodies: The Experimental Imperative in American Medicine," MacGovern Lecture presented at the annual meeting of American Osler Society, San Francisco, CA, 1996.

6. ACHRE, *Final Report* (Washington, DC, 1995). Also published as *The Human Radiation Experiments: Final Report of the President's Advisory Committee* (Oxford, 1996). Additionally, three supplemental volumes have been published by the U.S. Government Printing Office: I: Ancillary Materials (061-000-0850-1-9); II: Sources and Documents (061-000-0851-9); III: (061-000-0852-7). For an essay review of *Final Report* and a response from ACHRE's chair and executive director, see Robert Martensen, "If Only It Were So: Medical Physics, U.S. Human Radiation Experiments, and the *Final Report* of the President's Advisory Committee (ACHRE)," *Medical Humanities Review* 11.2 (1997): 21–37; Ruth Faden and Dan Guttman, "In Response: Speaking Truth to Historiography," *Medical Humanities Review* 11.2 (1997): 37–44.

7. JHL, quoted in JHL and Dan Wilkes, "History of the Donner Laboratory, Session Two," undated (JLP: #434-92-0066 / #19-14-6 / #2 / "S"): 5.

8. Claudia Clark, *Radium Girls: Women and Industrial Health Reform, 1910–1935* (Chapel Hill, NC, 1997). See also Robert S. Stone, "Health Protection Activities of the Plutonium Project," in *Symposium on Atomic Energy and Its Implications: Papers Read at the Joint Meeting of The American Philosophical Society and The National Academy of Sciences, November 16 and 17, 1945* (Philadelphia, 1946), 13–14; John B. Little, "Cellular, Molecular, and Carcinogenic Effects of Radiation," *Hematology and Oncology Clinics of North America* 7.2 (April 1993): 337–52.

9. JHL, quoted in JHL and Wilkes (n. 7 above), 3.

10. Stone, "Health Protection Activities," (n. 8 above), 13.

11. U.S. DOE, *Human Radiation Experiments: The Department of Energy Roadmap to the Story and the Records* (Springfield, VA, 1995): 19, 21. Update to Final Report*.

12. Ibid.

13. "More Irradiation of Americans Revealed: 9,000 People Were Exposed in Cold War Testing," *San Francisco Chronicle*, 10 February 1995: A7. Update to Final Report*.

14. DOE (n. 11 above), 200–270.

15. A. J. Lussenhop, J. C. Galfimore, W. H. Sweet, E. G. Struxness, and J. Robinson, "The Toxicity in Man of Hexavalent Uranium Following Intravenous Administration," *American Journal of Roentgenology* 79 (1958): 83–100; S. R. Bernard, "Maximum Permissible Amounts of Natural Uranium in the Body, Air, and Drinking Water Based on Human Experimental Data," *Health Physics* 1 (1958): 288–305.

16. Bertram V. A. Low-Beer, "External Therapeutic Use of Radioactive Phosphorus," *Radiology* 47.3 (1946): 213–22.

17. M. G. Rowley, D. R. Leach, G. A. Warner, C. G. Heller, "Effect of Graded Doses of Ionizing Radiation on the Human Testis," *Radiation Research* 59 (1974): 665–78.

18. G. V. LeRoy, J. R. Rust, R. J. Hasterlik, "The Consequences of Ingestion by Man of Real and Simulated Fallout," *Health Physics* 12 (1966): 449–73.

19. Lussenhop et al. (n. 15 above), 84; Bernard (n. 15 above), 299.

20. U.S. Congress, House of Representatives, Committee on Energy and Com-

merce, Subcommittee on Energy Conservation and Power of the Committee on Energy and Commerce, "American Nuclear Guinea Pigs: Three Decades of Radiation Experiments on U.S. Citizens" [hereafter Markey Report] (Washington, DC, 1986), 15.

21. LeRoy et al. (n. 18 above), 452, 473.

22. Several patients received doses known to be toxic. In Bernard's study on uranium toxicity, he notes that Lussenhop's study had found an uranium dose of 0.1 mg / kg to be toxic; this corresponds to 7 mg for a 70 kg male. The six patients in his study received doses ranging from 4 mg to 50 mg (Bernard, n. 15 above, 288–305).

23. Lussenhop et al. (n. 15 above), 84; Bernard (n. 15 above), 299.

24. Wright Langharn (Los Alamos Laboratory), in DOE (n. 11 above), 211.

25. Ibid.

26. Wright Langham (Los Alamos Laboratory), quoted in Philip J. Hilts, "Healthy People Secretly Poisoned in 4O's Test," *New York Times*, 19 January 1995.

27. Markey Report (n. 20 above), 15.

28. Henry K. Beecher, "Ethics and Clinical Research," *New England Journal of Medicine* (hereafter *NEJM*): 274.24 (1965): 1356.

29. Markey Report (n. 20 above), 1, 5.

30. Committee on Bioethical Issues of the Medical Society of the State of New York, "The Ethics of Using Scientific Data Obtained by Immoral Means," *New York State Journal of Medicine* 19.2 (1991): 58.

31. Susan E. Lederer, *Subjected to Science: Human Experimentation in America before the Second World* War (Baltimore, MD, 1995). Page references to this source are in the text.

32. George J. Annas and Michael A. Grodin, eds., *The Nazi Doctors and the Nuremberg Code* (New York, 1992), 2; Lederer (n. 31 above), 140.

33. DOE (n. 11 above), 24.

34. "Report of the UCSF Ad Hoc Fact Finding Committee on World War II Human Radiation Experiments," February 1995 (UCSF Special Collections Library [hereafter UCSFSCL]), 15.

35. Philip J. Hilts, "Inquiry Links Test Secrecy to a Cover-Up: A Fear of Publicity is Seen as a Motive," *New York Times*, 15 December 1994, A26.

36. Some work was published. Much remained as internal reports. Many studies did not disclose the source of their research subjects. Rowley's description of the testicular irradiation studies, for instance, makes no mention of the use of prison inmates (Rowley et al., n. 17 above, 665–78).

37. DOE (n. 11 above), 15.

38. Joseph Hamilton, "Letter to Robert G. Sproul (President, UC)," 5 January 1943 (JLP: #434-92-0066 / #19-14-6 / #2 / "S").

39. Raymond T. Birge, (Chair, Department of Physics, UCB), "Letter to Robert G. Sproul," 11 January 1945 (JLP: #434-92-0277 / #19-14-6 / #2 / "nuclear medicine").

40. "Minutes of the Regents (of UC)," 17 May 1946 (Bancroft Library: University Archives: Office of the President, Regents' Minutes, Committee on Finance and Business Management, vol. 46, 1946): 167, 411.

41. Birge, (n. 39 above).

42. "Joseph Gilbert Hamilton, Research Group Leader—A Biographical Sketch," undated (LBLARO: no file information given).

43. According to JHL, this was a "stunt." Hamilton knew that sodium would not concentrate in bone and that the dose was too low to be therapeutic. JHL, "Letter to Childs," 13 July 1966. See also [Hardin Jones], "History of Donner Laboratory," undated (Bancroft Library: University Archives: Hardin Jones: Committees, Speeches, Reprints, and Papers: file series #79/112C / carton #2 / folder "UCB, Donner Lab Histories & Report of Activities"): 5.

44. "Hamilton—Biographical Sketch," undated. For a variety of interpretations of his decision to proceed with the injection despite the orders, see: "Report of the UCSF ad hoc fact finding committee," 1995, 16–23, 21–22.

45. "Hamilton—Biographical Sketch," undated.

46. Hamilton, "Letter to Colonel K.D. Nichols," 31 December 1946 (LBLARO: Joseph G. Hamilton, Project 48A Reports / #19–14–35 / LBL-6215A / "Joseph Hamilton, Radioactive"). Page numbers in the text are to this source.

47. Department of Defense, Research and Development Board, Committee on Medical Sciences and Committee on Atomic Energy, Joint Panel on Medical Aspects of Atomic Warfare, "Program Guidance Report," 20 September 1951 (ACHRE: included as "Attachment 2" in memo "Human Experimentation in Connection with Atomic Bomb Tests," 8 September 1994).

48. "Project 4.5—Flash Blindness," undated (ACHRE: included as "Attachment 7" in memo "Human Experimentation in Connection with Atomic Bomb Tests," 8 September 1994).

49. Robert A. Hinners, (U.S. Navy), "Report on the Participation in Selected Volunteer Program of Desert Rock V-Z, 25 April 1953," 5 June 1953 (ACHRE: included as "Attachment 10" in memo "Human Experimentation in Connection with Atomic Bomb Tests," 8 September 1994). Excerpted as "Dispatch from Ground Zero," in Eileen Welsome, *The Plutonium Files: America's Secret Medical Experiments in the Cold War* (New York, 1999), 285–89.

50. "Memorandum for the Secretary of the Army, Secretary of the Navy, Secretary of the Air Force: The Use of Human Volunteers in Experimental Research," 26 February 1953, reproduced in Annas and Grodin, (n. 32 above), 343–45.

51. Shields Warren, "Draft Staff Paper on Troop Participation in Operation Tumbler-Snapper: Memo to Brigadier General K.E. Fields, Director, Division of Military Application," 25 March 1952 (ACHRE: included as "Document 4" in memo "Historical Background on U.S. Nuclear Testing," 7 September 1994).

52. General Cooney, in "Debate between Admiral Greaves, General Cooney, Dr. Warren, and others," c. 1950 (DOE Archives: file #326, collection "Div. Biology & Medicine." box #3218, folder "ACBM minutes": 6. Page numbers in the text are to this source.

53. JHL in JHL and Wilkes (n. 7 above) 7.

54. Ibid., 9.

55. Lussenhop et al. (n. 15 above), 83.

56. Bertram Low-Beer, "Radioactive Phosphorus as an External Therapeutic

Agent in Basal Cell Carcinoma, Warts, and Hemangioma," *American Journal of Roentgenology* 58.1 (July 1947): 4–9.

57. David A. Wood, "Historical Overview: Cancer Research Institute and Cancer Program, UCSF," undated (JLP: #434-92-0066 / #19-14-6 / # 2 / "W"); "The UC Program in Cancer Research," undated (Bancroft Library: University Archives: President Correspondence and Papers (PCP), ID #CU-5, carton #699, folder "140").

58. *NIH Factbook*, lst edition (Chicago, 1976). The NIH budget grew from $850,000 in 1946 to just over $4 million in 1947 and to $12,475,000 in 1948. In the same years, the National Cancer Institute budgets were $549,000, $1,821,000, and $14,500,000.

59. Michael B. Shimkin, "Lost Colony: Laboratory of Experimental Oncology, San Francisco, 1947–54: Historical Note," *Journal of the NCI* 60.2 (1978): 479–88.

60. "Protocol No. 2: Transplantation of Malignant Melanoma," 19 October 1951 (UCSFSCL: Goldfrey, Edwin Barkley, papers, #MSS88-56 / #6 / 25 / "Cancer Board 1951–52").

61. "Protocol No. 3," 5 June 1952 (EGP: #MSS88-56 / #6 / 25 / "Cancer Board 1951–52"). It seems strange to transfuse the irradiated patients with blood from other leukemic patients. In a different study, Shimkin explored the distribution of white blood cells in the body by "transfusing non-leukemic cancer patients with blood from leukemic donors" ("Laboratory of Experimental Oncology, Laguna Honda Home, S. F., California," 6 March 1950 (LBLARO: Scientists' Papers: Hardin Jones Papers: #19-14-5 / #1 / "application for cancer funds": 2).

62. "Cancer Board Meeting," 21 May 1952 (EGP: #MSS88-56 / #6 / 25 / "Cancer Board 1951–52").

63. Shimkin, (n. 59 above), 484.

64. Ibid., 487.

65. "Availability of Radioactive Isotopes: Announcement from Headquarters, Manhattan Project, Washington DC," *Science* 103.2685 (14 June 1946): 698.

66. Joseph Hamilton, "The Metabolism of Radioactive Elements Created by Nuclear Fission," *NEJM* 240.22 (1949): 863–70.

67. Joseph Hamilton, "The Rates of Absorption of the Radioactive Isotopes of Sodium Potassium, Chlorine, Bromine, and Iodine in Normal Human Subjects," *American Journal of Physiology* 12 (1938): 667.

68. LeRoy et al., (n. 18 above), 449–73.

69. R. L. Hayes, J. E. Carlton, W. R. Butler, "Radiation Dose to the Human Intestinal Tract from Internal Emitters," *Health Physics* 9 (1963): 915–20.

70. Markey Report, 1986: 13.

71. For example, see the following letters: A. R. Behnke (captain, U.S. Navy), "Letter to Dr. Will Siri," 22 January 1954; Behnke, "Letter to Dr. John Lawrence," 30 July 1954; Behnke, "Letter to Dr. Will Siri, 30 July 1954; all in JLP: #434-90A-0168 / #9 / "Kratzer material regarding India, AEC, Biology and Medicine-Pollycove."

72. J. M. Tanner, "Project: Body Structure," 8 November 1957 (LBLARO: Applied Science Division, William E. Siri, Scientists' Papers, 1945–1959 (WSP): ac-

cession #434-91-0131, file #19-14-18, carton #1/9, folder "Interoffice memo, 1945–1957")

73. Siri, "Letter to JHL," 10 January 1955 (WSP: #434-91-0131 / #19-14-18 / #1/9 / "Interoffice memo-1967"); Siri, "Letter to A. R. Behnke," 12 January 1955 (WSP: #434-91-0131 / #19-14-18 / #1/9 / "corres."). The archivists at the LBL investigated this project in 1994 and found no contemporary evidence proving that the injections of radioactive chromium ever took place. Existing documents only discuss planning the project. See "Research Fact Sheets: Topic #3: San Francisco 49ers and Cr_{51}," 13 December 1994 (LBLARO).

74. JHL, "Letter to Clinton Duffy (Warden, San Quentin Prison)," 29 January 1951 (JLP: #434-90-AO186 / #19-14-6 / # 4 / "R-Z 1951").

75. Ibid.

76. One published study only had 22 minorities ("Chinese, Filipino, North American Indian, Mexican, or Negro") out of 200 subjects. Does this reflect the inmate population, or the investigators' choices? (R. Wennesland, Ellen Brown, J. Hopper, J. L. Hodges, O. E. Guttentag, K. G. Scott, I. N. Tucker, and B. Bradley, "Red cell, plasma, and blood volume in healthy men measured by radiochromium (Cr_{51}) cell tagging and hematocrit: Influence of age, somatotype, and habits of physical activity on the variance after regression of volumes to height and weight combined," *Journal of Clinical Investigations* 38.7 (July 1959): 1065–77).

77. JHL, "Letter to Duffy," (n. 74 above).

78. James H. Corley (Comptroller, UCB), "Memo to JHL," 4 August 1949 (LBLARO: Administration Division, Business Manager, Research Development Administration Files: 1946–1957: accession #434-90-0020, file #13-11-14, folder "Medical Physics Gen. Correspondence").

79. JHL, "Letter to James H. Corley," 18 August 1949 (PCP: #CU-5 / 759 / "713"); L. I. Stanley (Chief Prison Medical Officer, San Quentin), "Letter to John Lawrence," 17 February 1949 (LBLARO: Administration Division, Business Manager, Research Development Administration Files: 1946–1957: accession #434-90-0020, file #13-11-14, folder "Medical Physics Gen. Correspondence").

80. JHL, and Donald Van Dyke, "Proposal for Erythropoiten Study," 30 March 1962 (JLP: #434-90-AO186 / #19-14-6 / #9 / "F, G, H").

81. JHL, "Letter to H. A. Gross (Chief Medical Officer, San Quentin)," 6 March 1962 (JLP: #434-90-AO186 / #19-14-6 / #9 / "F, G, H").

82. JHL, "Research Proposal sent to the Warden of San Quentin," 26 April 1962 (JLP: #434-90-AO186 / #19-14-6 / #9 / "F, G, H").

83. JHL, "Letter to Will Siri," 17 March 1952 (WSP: #434-91-0131 / #19-14-18 / #1/9 / "Interoffice memo, 1945–1957"). See also: Siri, "Fat, Water, and Lean Tissue Studies," undated (WSP: #434-91-0131 / #19-14-18 / #7/9 / "117–Abstracts Written").

84. JHL, "Letter to Will Siri," 4 August 1949 (WSP: #434-91-0131 / #19-14-18 / #1/9 / "Interoffice memo, 1945–1957"); Lawrence, "Project Description: Human Physiology and Experimental Medicine," 1 May 1958 (WSP: #434-91-0131 / #19-14-18 / #1/9 / "Interoffice memo, 1945–1957").

85. Pregnant women: W. J. Dieckmann (University of Chicago), "Letter to William Siri," 26 April 1956 (WSP: #434-91-0131 / #19-14-18 / #1/9 / "corres.").

Infants: Siri, "Letter to Theo C. Panos (University of Arkansas)," 12 January 1960 (WSP: #434-91-0131 / #19-14-18 / #1/9 / "corres."); Siri, "Letter to Samuel J. Fomon (State University of Arkansas)," 8 June 1960 (WSP: #434-91-0131 / #19-14-18 / #1/9 / "corres."). Premature Infants: B. M. Kagan (Cedars of Lebannon Hospital, Los Angeles), "Letter to Will Siri," 14 June 1957 (WSP: #434-91-0131 / #19-14-18 / #1/9 / "corres., 1945–1957").

86. "48 More Human Radiation Tests Revealed," *San Francisco Chronicle*. 28 June 28 1994, 4A; Paul Hoversten, "Fetuses, Stillborn Were Part of Radiation Tests," *USA Today*, 28 June 1994, 1A.

87. Charles Hayter, "Experience Not Experiment: The Introduction of Radium Therapy to Medicine," paper presented at New York Academy of Medicine Symposium: One Hundred Years of Radiosotopes and Health, 9 October 1998.

88. JHL and Wilkes (n. 7 above), 14.

89. A.V. Hill, quoted in JHL, "Isotopes and Nuclear Radiations in Medicine," (n. 2 above).

90. JHL in JHL and Wilkes (n. 7 above), 14.

91. We are indebted to Skuli Sigurdsson of the Max Planck Institute for the History of Science, Berlin, Germany, for directing us to Bohr. See also: F. Aaserud, *Redirecting Science: Niels Bohr, Philanthropy, and the Rise of Nuclear Physics* (Cambridge, 1990); Dominick Pestre, "The Decision-making Processes for the Main Particle Accelerators Built Throughout the World from the 1930s to the 1970s," *History and Technology* 9 (1992): 163–74. An important distinction may be made between the Lawrences (and their Berkeley colleagues) and Bohr and other European physicists, as well as East Coast physicists in the United States. In contrast to the dominant academic division of labor of doing high-energy physics from the 1930s forward, Ernest Lawrence and his colleagues self-consciously adopted an entrepreneurial ethos that tended to blur distinctions between theoreticians, experimentalists, and engineers.

92. Dan Wilkes, quoted in JHL and Wilkes (n. 7 above), 7.

93. Birge, "Letter to Sproul" (n. 39 above).

94. The Donner family provided generous support to Lawrence's work, starting in 1937 (JHL and Wilkes, n. 7 above, 1).

95. JHL, "Letter Robert G. Sproul (President, UC)," 1 March 1947 (PCP: #CU-5 / # 1182 / "Special Problem: Relationship Med. School and Div. Med. Physics").

96. [JHL or Hamilton], "Letter to Dean Smyth (UCSF)," 11 February 1949 (UCSFSCL: School of Medicine-Dean-Papers: #AR 90-56 / #6 / 13 / "organized research-metabolic unit-1955–prior").

97. William Kerr, in "Notes on a Conversation Between Drs. Kerr and Lawrence with Regard to the Proposed Metabolic Unit at Cowell," undated (JLP: #434-92–0066 / #19–14–6 / #18 / "Metabolic Unit Correspondence").

98. JHL, "Letter to Robert G. Sproul," 12 April 1950 (LBLARO: Life Sciences Division, R&D Administrative Files: #434-90-02116 / #16-5-39 / #4 / "Metabolic Unit Duplicates").

99. Kerr (n. 97 above).

100. Advisory Committee on Inter-Campus Medical Teaching and Research,

"Letter to Robert G. Sproul," 11 March 1952 (PCP: #CU--5 / #1182 / "Special Problem: Relationship Med. School and Div. Med. Physics"); Advisory Committee on Inter-Campus Medical Teaching and Research, "Letter to Sproul," 22 September 1953 (PCP: #CU-5 / #1182 / "Special Problem: Relationship Med. School and Div. Med. Physics").

101. In 1950, JHL had been appointed a Professor of Medical Physics at Berkeley. UCSF, however, citing his "lack of integration and cooperation," never promoted him past associate professor ("Brief of File on Promotion of JHL," 8 August 1950 (PCP: #CU-5 / # 1182 / "Special Problem: Relationship Med. School and Div. Med. Physics"). The Division of Medical Physics subsequently became its own department (JHL, "History of the Donner Laboratory," undated, n. 7 above).

102. "Availability of Radioactive Isotopes" (n. 65 above), 697–705.

103. Hamilton, "Letter to Paul C. Aebersold (U.S. Engineer Office, Research Division, Manhattan District)," 30 August 1946 (LBLARO: Ernest O. Lawrence general files, 1920–1959: #MSS 72 / 117c / 9 /18 / "26: Isotope Research").

104. "Use of Radioisotopes in Human Subjects" [letter fragment], 5 October 1949 (LBLARO: #434-90-0020 / #13-11-14 / "isotopes-gen. correspondence" / "8/26").

105. Jones, "Letter to Dr. S. Allan Lough (Assistant Chief, Isotopes Division, AEC)," 30 January 1952 (JLP: #434-90A-O168 / #19-14-6 / #4 / "AEC 1952").

106. JHL, "Memo to Members of the Advisory Committee (to the Division of Medical Physics," 10 April 1953 (LBLARO: file information illegible). By 1951, Berkeley had received "blanket authorization" (J. H. Gilette [superintendent, Radioisotopes Control Department, Oak Ridge National Laboratory], "Letter to UC Radiation Laboratory," 1951 [LBLARO: Donald Crocker, Laboratory Associate Director's Files: #434-90-0020 / #13-11-14 / MU-5 / "Chemistry Program-Isotope Procurement"]).

107. "Cooperative Work with Other Research Groups outside of Donner Laboratory," undated (LBLARO: Life Sciences Division, R&D Administrative Files: #434-90-02116 / #16-5-31 / #3 / "Donner Historic Personnel Listing").

108. Myron Pollycove, "Letter to the AEC: Report on Foreign Travel," 1962 (JPL: #434-90A-O168 / #19-14-6 / # 9 / "Kratzer material regarding India, AEC, Biology and Medicine-Pollycove").

109. JHL, "Memo to Members of the Advisory Committee" (n. 106 above).

CHAPTER FIVE

"I Have Been on Tenterhooks"

Wartime Medical Research Council Jaundice
Committee Experiments

Jenny Stanton

Certain experiments become notorious in the history of medicine or medical ethics, while other quite similar episodes remain in obscurity. Thus, Saul Krugman's long series of trials of hepatitis transmission (1956–71), using inmates of a home for mentally retarded children in New York, fell under the spotlight when Henry Beecher wrote about them among other examples in his 1966 essay on medical experiments.[1] They remained a matter of controversy thenceforth. Yet there has been no equivalent discussion of comparable (though smaller-scale) experiments conducted in Britain during World War II, using conscientious objectors and hospital patients as subjects. Obviously, this is partly because the wartime experiments were cloaked in secrecy. Their outcomes were reported in readily accessible medical journals, with a less accessible but much fuller account given in the Medical Research Council (MRC) report series, but the details of the experiments were not public in the way Krugman's were. Some would argue that there was a definite shift in ethical perceptions between the war years and the 1960s, partly due to revelations about what occurred during the war in Germany rather than in Allied countries.[2] But others warn that we need a better understanding of the construction of "medical ethics" if we are to avoid reading backward and seeing what was done and how it was judged in terms of current debates.[3]

This chapter is not directly about medical ethics. It aims, first, to unravel the story of these U.K. experiments, and second, to understand what motivated the people involved in backing and conducting them and how they responded to the pressures their actions generated. It does this through looking at the published outcomes; through an interview with the chief investigator, Fred MacCallum, nearly fifty years after the events; and above all, through the examination of files in the MRC archives relating to the Jaundice Committee during 1942–47, when the experiments and allied studies were conducted. There are many omissions, but the intention is to provide sufficient record to allow for discussion of the issues raised by these experiments.[4]

I begin by outlining MacCallum's prewar involvement in virology, and especially yellow fever vaccine development, which fed into wartime work on hepatitis, because the vaccine sometimes caused large-scale outbreaks of jaundice among troops. There appeared to contemporary observers to be many forms or settings of infectious jaundice as well as a bewildering and proliferating array of names for supposed variants, some of which appear in this essay. Hepatitis outbreaks among troops were the immediate reason for the establishment of the Jaundice Committee, whose activities are chronicled here. I follow with a short discussion of Krugman's experiments, for comparison. The quotation in the title of this paper comes from a comment—a sort of typewritten sigh of relief—from the very senior clinician who chaired the wartime Jaundice Committee. In context, it reveals the extent of the qualms felt at the time by some who bore responsibility for the experiments, though not perhaps by all those actually conducting them.

Background

Jaundice, as an illness in which the liver is involved and the patient turns yellow, has been observed since ancient times. In the nineteenth century when the new chemical industries produced a spate of toxin-induced liver disorders, experts in Germany argued that acute jaundice was mainly noninfectious. However, this view was challenged at the end of the century by outbreaks in military campaigns and other settings where large numbers of people were crowded together. By the early twentieth century, "infectious jaundice" was becoming an established notion.[5] By the 1930s, some researchers were investigating the idea that it was caused by a virus—although the concept of the virus was mainly negative at that point, as an infective agent that was nonfilterable, that is, much smaller than a bacterium.[6]

Papers published in 1937 and 1939 by G. M. Findlay and F. O. Mac-Callum exemplify the way virology often worked at that time, by a process of elimination.[7] They pointed out that no micro-organism had been found for hepatitis that could be seen under a microscope or trapped in a filter and cultured; therefore, the causal agent could be presumed to be a virus. Thus, on the threshold of the war, hepatitis was one among many diseases with candidate viral etiology. But with the limited scientific tools at their disposal, it was not obvious how these virologists could hope to take matters further. In fact, they were concentrating on rather different problems that were to bring them back to hepatitis in an unexpected way.

According to his own account, Fred MacCallum left Toronto for the United Kingdom in 1934 because he wanted to learn more about viruses, and resources were poor in Canada after the recession.[8] In Britain, he knew of three centers currently studying viruses (among other micro-organisms): the Lister Institute; the Medical Research Council unit at Hampstead; and the Wellcome Bureau of Scientific Research on the Euston Road, where the whole of the fourth floor was taken up with tropical medical research. Half the area was occupied by chemists working on antimalarials and leishmaniasis; while Findlay worked alone on yellow fever, lymphogranuloma, and rift valley fever. The Euston Road laboratories were the nexus of a chain reaching into Africa, with Wellcome's research laboratories in Khartoum and its mobile, floating laboratory on the Nile, and connecting with the laboratories at Beckenham in Kent and others in the United States, where drugs were manufactured.[9]

MacCallum originally took a job at the London Hospital under Professor S. P. Bedson, the bacteriologist who had discovered psittacosis, but in July 1936 he secured a post as assistant to Findlay for research on yellow fever vaccine. Their Wellcome salaries were subsidized by the Colonial Office, which was concerned about yellow fever as a scourge of white officials and traders in West Africa. Others had long been searching for a vaccine, notably the Americans in relation to the building of the Panama Canal;[10] and Findlay collaborated with both the London School of Hygiene and Tropical Medicine (LSHTM) and the Rockefeller Foundation in New York, which helped support the U.S. interests in public health in South America.

Findlay and MacCallum produced a live-virus yellow fever vaccine, manufactured from the brains of mice inoculated with infected serum. Passage through mouse brains was thought to attenuate the virus partially, but not sufficiently for complete safety; therefore, large doses of serum from convalescent yellow fever patients (convalescent serum) were added to the freeze-dried mouse brain extract to counteract any remaining virulence.[11] Later, the linked research teams managed to grow the virus in chick embryos, which provided a more satisfactory, more controllable medium. Rockefeller researchers found that after many passages through chick embryos the virus vaccine was so attenuated that no anti-serum need be added; this vaccine proved satisfactory in trials conducted in Brazil in the 1940s.[12] However, although yellow fever vaccine had few precedents—it was the first virus vaccine for humans after smallpox and rabies—fixed ideas seem to have developed around it rather rapidly. The notion that the freeze-dried vaccine must be made up with serum rather than water persisted throughout the Second World War. As larger amounts were needed, normal serum replaced convalescent serum.

Soon after MacCallum joined Findlay, cases of jaundice began to occur in people who had received yellow fever vaccine before going to Africa. MacCallum, accustomed to notions of noninfectious jaundice, made little of this. Findlay was more concerned, and he began to wonder about connections between this and other instances where jaundice followed injections. Together with MacCallum, he embarked on a literature search that revealed occasional cases of jaundice in various types of clinics—diabetes, arthritis, and others—in many different countries. They published on this phenomenon in 1937,[13] stating that their cases could not be yellow fever— the apparently obvious explanation in view of the coloration—since the jaundice occurred about sixty days after receiving the vaccine. Moreover, blood samples taken from some patients established that they had developed antibodies to yellow fever ten to fourteen days after inoculation. The findings pointed to hepatitis of a type analogous to that observed in the clinic cases they had surveyed.

Following this episode of 1936–37, Findlay and MacCallum had a clear period of about five years during which they supplied yellow fever vaccines without further cases of jaundice. At the outbreak of the war, Findlay was sent to West Africa as a tropical disease adviser, and MacCallum was left making yellow fever vaccine "with a couple of technicians." He was required to increase production from some twenty milliliters per week to several thousand milliliters to provide for all service personnel going to West Africa.[14] Besides needing to increase the output of mouse brains, he re-quired—according to the then-accepted practice—an enormously in-creased supply of serum for dilution of the vaccine. With apparent good fortune, a newly developed technology was available to channel large vol-umes of serum in a compact form: freeze-dried plasma, using pooled serum derived from many donors. The newly organized Blood Transfusion Service, set up in 1938–39 in expectation of the war, relied heavily on freeze drying of blood and plasma, manufactured with the participation of the Wellcome Foundation.

Thus, when MacCallum was asked to step up yellow fever vaccine pro-duction, he called on a Wellcome contact at Beckenham and secured a bot-tle of freeze dried plasma, which he reconstituted with water and incor-porated into a batch of vaccine. Three months later, he was telephoned by the director of the Royal Air Force medical services, who had suffered a nasty attack of jaundice sixty-six days after yellow fever inoculation.[15] This was one of several cases, the most severe in terms of the seniority of the victim. MacCallum, knowing the batch number of the dried serum he had used, telephoned his Wellcome contact, who still had some bottles of the same batch of plasma in store. These were used in some of the exper-iments that followed, instigated by the War Office.

The Wartime Jaundice Committee and Research Team

As the war progressed, hepatitis became a cause of concern on a num-ber of fronts. On the one hand, there were outbreaks of jaundice among troops stationed in the North African desert and in Italy—about 16,000 cases, with a few deaths, between 1941 and 1943, mostly ascribed to "in-fectious hepatitis."[16] On the other hand, outbreaks associated with yellow fever vaccine were an increasingly serious embarrassment for the War Of-fice. In the past, yellow fever had been a major impediment in the prose-cution of military ventures in the Tropics;[17] so now the vaccine was seen as an essential safeguard. When this prophylactic measure turned troops yellow, mimicking the disease it was designed to prevent, it mocked the progress of British tropical medicine.[18] With America joining the war, there was a most spectacular vaccine-associated jaundice outbreak: 28,000 American troops were affected in the first six months of 1942 (with 62 deaths), following inoculation with yellow fever vaccine made by the Rockefeller group, evidently still using serum to dilute the attenuated vaccine. These thousands of cases presented a frightening specter of medically induced mass disablement.[19] When 500 American troops, newly arrived in Northern Ireland in 1942, suffered jaundice, the British became alarmed over the possible spread to the civilian population.

As so often happened, the impact of war—in this case, an indirect impact via a preventive health measure—acted as a stimulus to action on a medical front. It appears that the British and the U.S. military agreed that research was needed to stem the flood of jaundice cases. Although there was an understanding that the U.S. Army was to investigate the yellow fever association and the British to concentrate on infective hepatitis, in fact researchers on both sides of the Atlantic looked at every variant of hepatitis, since so little was known about it.

Sir Wilson Jameson, Chief Medical Officer at the Ministry of Health, asked the MRC to correlate existing research on jaundice and coordinate further investigations. A joint committee was established with MRC, armed forces, and Ministry of Health representation. This Jaundice Committee met six times between March 1943 and May 1945, with a postscript gathering in October 1945 to settle its affairs. Clearly, many negotiations were conducted before and between meetings. In fact, at the first meeting, at which a research team was selected, names had already been agreed upon, the only proviso being that the Wellcome Research Institute would have to be asked to release MacCallum for this work. A laboratory in the Department of Pathology at Cambridge was allocated for use by the research team, probably thanks to connections of one of the committee members.[20] The Ministry of Health was to make jaundice notifiable in Civil Defence Region 4 (East Anglia and adjoining counties) to allow epidemiological surveillance of a normal civilian population of some two and a half million. All cases of jaundice among troops stationed in the area were to be closely monitored by the research team. Already by the first meeting, the use of "human volunteers" for transmission experiments was under discussion.

The chair of the Jaundice Committee was Leslie Witts, Nuffield Professor of Medicine at Oxford. Witts and Edward Mellanby, secretary of the MRC, probably guided the selection of committee members, though the other joint bodies (the armed forces and the Ministry of Health) put forward their own men. A note from Witts to Mellanby late in 1943 reveals something of the personal element that must often have played a part in the selection process: "Poole [Major-General L. T. Poole, a medical supremo at the War Office, already on the Jaundice Committee] is very anxious that Biggam [another medical major-general at the War Office] should be invited to become a member of the Jaundice Committee. Biggam is taking an active part in the army's jaundice research and he is a person with whom I very much like working."[21] Mellanby made sure that Biggam was invited. Members, besides representing interested bodies— the army, the Ministry of Health, and the War Office—had to be eminent persons, known to the initiators, and, it would seem, compatible with the chairman. Almost all were London-based except Witts in Oxford, W. J. Tulloch, professor of bacteriology at St. Andrews in Scotland, and A. M. McFarlan, an epidemiologist at the Emergency Public Health Laboratory at Cambridge who acted as secretary to the Jaundice Committee and was also a member of the research team.

Each of the five members of the research team covered a particular aspect. McFarlan, seconded from the Public Health Laboratory Service (PHLS), conducted the epidemiological surveys. Clifford Wilson, a senior army physician, undertook clinical observations. M. R. Pollock, a bacteriologist from the PHLS, dealt with the biochemical problems of early detection of infective hepatitis (prior to onset of jaundice) and assessment of liver function in relation to different treatments. J. A. R. Miles, a clinical pathologist from the army, conducted hematological and serological investigations. Transmission experiments—the main concern of this account—were the responsibility of MacCallum.

It appears from the final report of the research team that MacCallum initially concentrated on finding an animal model; but in fact, he had already done this work before the Jaundice Committee was established, as he reported to an MRC subcommittee on "Jaundice in Industry" in November 1942.[22] Noting the almost totally unsuccessful work of other researchers, he had tried pigs, golden hamsters, Orkney voles, cotton rats, guinea pigs, canaries, mice, and rats, all with negative results. The failure of these earlier attempts to find an animal in which hepatitis could be produced led the Jaundice Committee to support MacCallum's call to use human beings as experimental subjects: "It was decided, therefore, that experiments on human volunteers were essential if further knowledge was to be obtained on the mode of spread and duration of infectivity of the various types of hepatitis designated as infectious, homologous serum and arsenotherapy hepatitis."[23]

How were volunteers obtained? The first line was to try conscientious objectors. MacCallum "went to talk to Quakers in that building [the Friends' Meeting House] next to the Wellcome Institute on the Euston Road" and persuaded "first one then another" to participate. This source was not plentiful enough, however. Soon "there weren't any more conchies" willing to act as experimental subjects.[24] Dr W. H. Bradley, a Ministry of Health appointee on the Jaundice Committee, suggested that rheumatoid arthritis patients might be recruited on the basis of reports in the prewar literature suggesting an attack of jaundice sometimes brought about remission of arthritis symptoms.[25]

Witts leaned on rheumatology colleagues and secured a group of volunteers in a unit in London for what was billed as a therapeutic trial of the effects of jaundice on rheumatoid arthritis. Witts used the term "inoculation" of the procedure used in these trials, but MacCallum describes various methods of administering the infected material—nasopharyngeal washings, blood, urine or feces from hepatitis patients—including spraying into the nose and mouth, swallowing, and injection. In the case of feces, which MacCallum left till last, a suspension in orangeade was apparently most favored among the recipients.[26] Infective material was derived from Wilson's patients, mostly service personnel in East Anglia (for infectious hepatitis); and from cases of post-transfusion hepatitis supplied by the Blood Transfusion Service.[27]

By March 1944 a fresh supply of volunteers was needed for further transmission experiments. The Jaundice Committee decided to request the use of military prisoners both in the Middle East and in the United Kingdom as well as civilian prisoners. Witts asked Mellanby to contact the civil and army authorities and provided him with a persuasive case, including statistics that appear, in retrospect, rather chilling: "The risk of fatality is probably no greater than is represented by a fatality rate of 8 in 10,000 cases in the recent epidemics in the Middle East. The risk of subsequent disability is probably about 1 per cent of cases. These rates of mortality and disability apply to individuals actually contracting infective hepatitis, and these would be only a small fraction of the total number of volunteers inoculated."[28] Witts speculated that men who were serving sentences for desertion or cowardice might "welcome this means of rehabilitating themselves in the eyes of society," that civilian prisoners would like to contribute to the war effort, and that all would welcome remission of their sentences.

But the prisoners were never subjected to this tempting offer, since the

adjutant-general ruled that the need for six months observation of experimental subjects might hamper remission for military prisoners who were
in for short sentences. Besides, as Lieutenant-General Sir Alexander Hood,
director-general of the Royal Army Medical Corps added, in relaying the
decision to Mellanby: "Though the risk of fatality is exceedingly low, there
might well be a death in the earlier stages of the experiments, and this
might easily lead to very considerable trouble."[29] Mellanby drew a similar
blank with his request to the prison commissioner for the use of civilian
prisoners, on the grounds that additional remission (above that normally
allowed for good behavior) would not be acceptable to the Home Office.[30]

Refused the use of prisoners, and seeing problems with other possible
groups that they discussed (such as inmates of lunatic asylums and monas-
tic orders), the Jaundice Committee pressed ahead with a search for further pools of rheumatoid arthritis patients. A letter to the *Lancet* was drawn
up over the signatures of Bradley and MacCallum on the beneficial effects
such patients sometimes experienced with jaundice. Before this public assertion of benefit could be made, however, the Jaundice Committee had to
assuage official concerns about potential damage to the health of volunteers. As Witts confided to Landsborough Thomson, second secretary to
the MRC, in July 1944, one of Bradley's superiors at the Ministry of Health
was "very worried about his connection with this work and raised very
strong objections to publication unless it had the declared support of the
Council."[31]

The requisite support for publication was nevertheless secured, with the
assurance that the Jaundice Committee fully recommended it. Their
grounds for so doing were partly that transmission experiments had already
shown that the feces of patients with infective hepatitis contained an infectious agent—a finding of great practical importance—and partly the desire to establish Bradley's priority with regard to this transmission and the
use of the infectious agent in treating rheumatoid arthritis patients. After
this, Bradley was no longer to be closely associated with transmission experiments.[32]

The Ministry of Health had reason to be wary of their man's name
being associated with further experiments. The ministry itself had requested the Jaundice Committee to look into what was termed "homologous serum jaundice" in the context of transfusions of blood and serum.
Bradley told a Jaundice Committee meeting in July 1944 that "the Ministry of Health had records of 200 cases of hepatitis in transfused persons

with 5 deaths. . . . The Ministry was concerned about the possibility of public clamour if it became known that many cases of jaundice and some fatalities were due to transfusion."[33] Most of these cases had been reported by doctors J. F. Loutit and Janet Vaughan, both of whom attended this meeting.[34] Vaughan requested a full-time social worker to assist her search for further information on the links between blood transfusion and jaundice, which the MRC agreed to fund.

Although the Ministry of Health was worried about an outcry if it were publicly known that transfusion might be associated with jaundice, they required further understanding of this type of transmission to amplify the tentative findings reported by MacCallum and Bradley.[35] To those involved with the transmission experiments, it was clear that further trials would concentrate on serum jaundice, rather than on the less harmful infectious jaundice. A third variant was added, representing what MacCallum referred to as the "social aspects" of the disease;[36] that is, jaundice associated with the arsenical treatment of venereal diseases: arsenotherapy or arsphenamine jaundice. The theory that this type of jaundice arose as a side-effect of arsenical drugs had survived from observations during the First World War until well into the 1940s, but more recently the prevalence of jaundice in venereal disease clinics attended by Italian prisoners of war had worried the military doctors. The clustering of cases suggested that an infectious agent might be responsible—something that was inadvertently transmitted via needles and syringes.

By September 1944, Witts had taken steps to facilitate the new round of experiments. As he told Mellanby: "We have provided 58 patients with rheumatoid arthritis for MacCallum to inoculate here in Oxford. Although I say 'we,' my Assistant Director, Dr. Alice Stewart has made all the arrangements."[37] These arrangements included the opportunity to draw on patients at another center: as Witts explained, "Dr. Alice Stewart is the daughter of Naish, Emeritus Professor of Medicine at Sheffield, and she has a number of connections there. We have made tentative enquiries and it would be possible for us to work up the Sheffield area and collect at least 100 volunteers with arthritis, probably more."[38] Oxford and Sheffield were the main centers for the expanded human transmission experiments, but other cases were "made available" in hospitals in Scotland, Wales, and elsewhere in England.

Sensitivity on the part of hospital authorities to the possible public view of these transmission experiments emerges in a couple of instances. The medical superintendent of Aberdeen Royal Infirmary, Dr. J. C. Knox, wrote directly to Mellanby, pointing out that a voluntary hospital that was: "very dependent on public trust and goodwill for its financial support cannot afford any suggestion that patients, even volunteer patients, are being "experimented" upon."[39] Mellanby, after referring the question to members of the research team in Cambridge (MacCallum and McFarlan), assured Knox that the risk to his arthritis patients from jaundice therapy was no greater than with gold therapy, a more favored experimental treatment, and advised that he emphasize this therapeutic aspect to the hospital's board of governors. But Mellanby also stressed the question of the national interest: jaundice research was "of high priority in relation to the war."[40]

MacCallum's typical weekly schedule during the peak period of transmission experiments was fairly hectic.[41] He spent Monday mornings at the Wellcome Bureau laboratories in the Euston Road, where he was still making yellow fever vaccine, with a conscientious objector as assistant. On Monday afternoons he went up to the headquarters of the jaundice research group in Cambridge to coordinate the team's work (and perhaps collect clinical material). He would spend Tuesday and Wednesday in Sheffield conducting transmission experiments on volunteer arthritis patients and then return to Cambridge on Thursday to monitor the progress of his animal experiments. Friday would be spent back in London. Meanwhile, McFarlan was working in East Anglia, looking at outbreaks of hepatitis in schools, nurseries, and a large institution for mental defectives where there were eighty-five cases of "infective hepatitis" in an outbreak in 1944.[42] Wilson was in Cambridge making clinical observations on patients—three hundred of the two thousand servicemen in the region notified as cases of infective hepatitis—while Pollock developed biochemical tests for early detection of infective hepatitis and changes in liver function. Miles, also at the jaundice research team's Cambridge headquarters, worked on hematological and serological reactions to clarify the clinical profile and distinguish the different types of hepatitis.

Toward the end of the war (and of the Jaundice Committee's activities), there was cause for further alarm over potential objections to the transmission experiments, and this time there is evidence of deliberate evasion. MacCallum had been publishing his findings in a series of articles, each of which had first to be submitted to the Jaundice Committee for approval. The last in the series dealt with arsenotherapy jaundice, which appeared to be transmitted by blood, but not by feces and nasal washings. This was an important finding—but there was a problem, as MacCallum had to confess to Mellanby: "I had included Dr. Alice Stewart's name, as we had done the work together, but as you will see she has erased this, as she felt the situation in Sheffield would be happier if the clinic did not realise that material from patients receiving arsenotherapy had been inoculated into their patients."[43] The real problem, which MacCallum avoided spelling out, was a fear of possible syphilis transmission alongside hepatitis, since arsenic therapy was used for treatment of syphilis. MacCallum was confident that his methods ensured that the material he used would carry only hepatitis, not syphilis, but this part of the trials was potentially very controversial.

MacCallum's experiments using material from two patients who had become jaundiced during arsenical treatment, with nineteen volunteers as recipients, confirmed the view that an infective agent carried from one patient to another via needles and syringes rather than the arsenic itself might be responsible for so-called arsenotherapy jaundice. The infective agent appeared to be the same as for serum hepatitis (as in the cases of vaccine and transfusion hepatitis). There were indications that better sterilization of needles and syringes could stop transmission.[44] These were important findings. But Witts well understood Alice Stewart's refusal to associate her name with any publication arising from these experiments. As he told Mellanby:

> I have been on tenterhooks about this work, as it has been carried out in patients with rheumatoid arthritis under the guise of homologous serum jaundice. . . . I have become increasingly uneasy about the issue raised [of possible transmission of syphilis]. . . . At the meeting in December . . . I got the Jaundice Committee to give a ruling that experiments on the transmission of post-arsenical jaundice must not be carried out on patients with rheumatoid arthritis, and I believe that no further experiments of this kind have been performed since that date. . . . I am hopeful that this is the last hurdle which the Jaundice Team faces. I must confess that this study of human transmission has caused me a good deal of worry and it is a great relief that no permanent ill effects have been observed in any of our volunteers.[45]

Whether by "permanent ill effects" he meant syphilis or hepatitis, Witts could indeed count himself lucky that no patients showed lasting damage—and that there were no deaths from the more serious serum hepatitis.[46] Mellanby, acceding to Witts's request for clearance of MacCallum's article, without recourse to the Jaundice Committee, commented that "publication certainly has my approval and, although some people might regard it as strong meat, I realise that it is the kind of work that had to be done."[47] Shortly afterward, MacCallum was moved to typhus research and the Cambridge jaundice team was dispersed.

There was then a rather inexplicable gap. The final report of the research team was published in 1951, some four years after the last recordings were made. Its preamble stated that the MRC had decided not to prioritize publication because many of the findings had been published during the course of the investigations and other reports now took first turn.[48] But this seems an inadequate explanation; perhaps the delay was occasioned, at least in part, by the nervousness so eloquently displayed in the Jaundice Committee files.

The major distinction between two types of hepatitis does not stand out among the wide-ranging series of conclusions in the MRC report, due to the broad scope of the inquiries and the various prior classifications of types of jaundice. For instance, under the summary findings for serology: "The evidence suggested a different causation for infective hepatitis from that of homologous serum or arsenotherapy hepatitis, but there is no evidence for or against cross-immunity between the latter two conditions."[49] In discussion of transmission experiments, a new term was introduced: "Infective hepatitis is believed to be due to a virus called virus A . . . Virus B causing homologous serum hepatitis has not been found in faeces . . . the derivation of virus B and its possible relation to virus A remains undecided."[50] Would it be hindsight to see the distinction between hepatitis A and B as the most important achievement of the Jaundice Team? The same point was stressed in the MRC's preface to the report, without using the terms A and B: "The outstanding findings of the human experiments were that a virus is present in the blood in arsenotherapy jaundice and that virus is excreted in the faeces in infective hepatitis."[51]

A wider audience had probably been reached through a short piece published anonymously in the *Lancet* in 1947, almost certainly written by MacCallum.[52] Here, the proposition was made for the first time to use the term *hepatitis A* for the short-incubation (20–40 days) form previously known as "catarrhal jaundice" or infectious or infective hepatitis; and to use the term *hepatitis B* for the long-incubation (60–100 days) form, homologous serum jaundice (though the incubation periods were not spelled out here). This clear statement of two different types of hepatitis with two different routes of transmission thus reached a public forum four years before the MRC report was published.

Much remained a mystery regarding the nature of hepatitis B, and its epidemiology was far more obscure than that of hepatitis A. MacCallum thought the high attack rate among his experimental subjects, inoculated with B, suggested that "only a small proportion of the population has been exposed to this agent as compared to virus A in England."[53] Perhaps this was a new disease, or perhaps natural transmission was extremely difficult. The apparent increase in cases over the years, MacCallum speculated, might be due to better recognition, or to an actual increase due to the more widespread use of blood products. There was even a possibility that the viruses interacted, since some evidence hinted that individuals who had recovered from B were more susceptible to A than normal.

We can see in this sort of speculation an image of hepatitis B as a rare, possibly new disease, chiefly associated with medical procedures involving blood, serum, or plasma. The virus had been found to be tough, yet the disease apparently failed to spread widely where there was no puncture of the skin by needle. It was not transmitted by the fecal-oral route like hepatitis A, or by droplet infection like so many other infectious diseases. Yet it sometimes appeared in clusters. Buried in the MRC report, in McFarlan's discussion of two outbreaks of hepatitis in a mental institution, was a possible clue to this clustering: he referred to both types of hepatitis (the prior outbreak was supposed to be B, and the one he studied in 1944 to be A) as having spread partly through "contact."[54] While McFarlan emphasized the uncleanly habits of the "low-grade defectives" in relation to the spread of infective hepatitis,[55] the pattern of spread among people living in close proximity echoed that observed in families in the villages the team had studied.[56]

Postwar studies in a mental institution in the United States were to further elucidate the nature of the transmission routes, including the meaning of "contact." One of the striking things about these postwar experiments is the extent to which they replicated the British wartime experiments.[57] It is likely that Krugman and his colleagues had not read the MRC report.

Postwar Experiments on Hepatitis: Krugman's Willowbrook Studies

The hepatitis experiments carried out by Saul Krugman between 1956 and 1971 subsequently received both high commendation and (to a much

greater extent) deep opprobrium; but at the time they started, they appear
to have been fairly uncontroversial. Krugman was a New York pediatrician
with a position, from 1946 onward, in the Department of Pediatrics at New
York University, where he worked with colleagues on infectious diseases
of children, particularly measles and rubella. His interest expanded to hepatitis; and in 1956, together with Joan Giles and Jack Hammond, he began
a series of studies in Willowbrook, a residential school on Staten Island in
New York housing about five thousand mentally defective children between the ages of three and ten years old. Within this institution—as in
many such institutions for the mentally deficient—viral hepatitis appeared
to be common. Krugman's team investigated the types of hepatitis involved
and the means of transmission by administering infective material to newly
admitted children. A special hepatitis unit was established in the school,
and those children whose parents agreed to submit them to the trials were
given fecal material or serum from hepatitis sufferers, either in drinks or
by injection.[58]

Fifteen years of experiments on several hundred children at Willowbrook resulted in many papers published in leading American medical
journals and wide acclaim for Krugman's achievements.[59] However, the
Willowbrook trials came under increasingly hostile scrutiny, and in 1966
Henry K. Beecher, anesthesia professor at Harvard, included Willowbrook
among twenty-two studies whose ethics he questioned, in an article on the
ethics of clinical research.[60] Criticism of the experiments from an ethical
standpoint continued over many years. However, many of Krugman's colleagues in the hepatitis world stood by him, organizing the second international symposium on viral hepatitis in 1981 as a tribute to Krugman,[61]
defending his experimental protocols and "assurance of truly informed
consent."[62]

Following Beecher's 1966 article, the *Journal of the American Medical Association* continued to voice support for Krugman, alongside further Willowbrook papers—at the same time as giving a favorable review of
Beecher's work on medical ethics.[63] The debate over Willowbrook spilled
over into the British journals, which, like the American medical press,
tended to be impressed by Krugman's work. In 1971 Stephen Goldby, a
doctor at the Radcliffe Infirmary in Oxford, wrote to the *Lancet* asking if
it could be right to perform an experiment on a child when no benefit
would result to that individual; in his view, the answer must be no.[64] Although the *Lancet* printed replies from Krugman himself and from other

doctors—including Pasamanick of the New York Department of Mental Hygiene—who were involved in the Willowbrook project, its subsequent editorial policy was critical of the experiments.[65]

What had Krugman achieved with the Willowbrook experiments? Foremost was the distinction between two types of hepatitis, which he labeled MSI and MSII, corresponding to A and B; the first having a fecal-oral route of transmission and a shorter incubation period, and the second a mainly parenteral route of transmission and a longer incubation period. There was some suggestion that MSII was transmissible by mouth, but to a lesser degree. C. M. McKee, a British public health doctor, in an historical review in 1988 noting the previously accumulating evidence for two distinct types of viral hepatitis, states: "The existence of separate hepatitis A and B viruses was finally confirmed by Krugman in the Willowbrook experiments."[66] McKee cites MacCallum's 1947 paper but not the 1951 MRC report that covers the British wartime hepatitis studies in full detail.

Krugman himself, in a 1978 overview paper, cites the MacCallum and Bradley letter of 1944, together with five other human transmission studies from the 1940s, as precedents for his own work; he omits the 1951 MRC report.[67] His own interest, according to this account, was sparked by a symposium on laboratory work on hepatitis sponsored by the National Academy of Sciences National Research Council and the Armed Forces Epidemiological Board at New York University and Bellevue Hospital in 1954. The comprehensive failure to propagate hepatitis in laboratory animals pointed to the necessity for further human experiments. Krugman does not here mention that his subsequent research was partly funded by the United States Armed Forces; according to the historian William Muraskin, the army was "the major sponsor" of Krugman's Willowbrook work.[68] There would appear to be a continuity in military interest in hepatitis research from the 1940s into the postwar period.

In his 1978 account, Krugman included among his summarized results the observation "that HB could be spread from person to person following the type of prolonged, intimate contact that involved sharing of excretions. Thus, it was clear that a parenteral [e.g., inoculation] type of exposure was not the only mode of transmission of HB infection."[69] Although Krugman's group had published on the possibility of oral transmission of MSII, the singling out of "intimate contact" here seems a post hoc recognition of an important facet that really only became clear to clinicians during the 1970s and was not originally picked up by Krugman:

that is, sexual transmission. Later, in the 1980s, this route of transmission came to be seen as even more important, with hepatitis B serving as an epidemiological model for AIDS.

The other outcome of the Willowbrook experiments—apart from the confirmation of two types of hepatitis—was the preparation of a crude vaccine by boiling serum containing the hepatitis B virus. MacCallum had also attempted to inactivate the virus, in order to make the serum for transfusion safe, rather than to prepare a vaccine.[70] In this sense, Krugman was definitely taking a step further than his predecessors, but his vaccine was too experimental to be tried outside Willowbrook, where it was only used on a small group of children with mixed results. The later development of an effective vaccine depended heavily on the discovery of the hepatitis B antigen (referred to at the time as the "Australia antigen") in the late 1960s by Baruch Blumberg in Philadelphia.[71]

There is a further dimension to the vaccine story at Willowbrook. Early in his investigations there, Krugman had managed to reduce hepatitis by some 80 percent by administering gamma globulin, an established "passive vaccination" prophylaxis for hepatitis.[72] Krugman's work on the active vaccine was subsequently emphasized at the expense of the immunoglobulin findings. Yet from the viewpoint of inmates and staff, his use of gamma globulin was more effective. While his transmission experiments were continually justified on the grounds that most children admitted to Willowbrook were bound to catch hepatitis, Krugman's own work with passive vaccination showed this need not be the case.

A final aspect of the Willowbrook work should be considered as a partial explanation for the support for Krugman among his colleagues. Krugman recorded that "many thousands of serum specimens collected over a period of about 20 years have been stored in a 'serum bank.' These valuable sera were obtained before, during and for many months and years after onset of HA and HB. These pedigreed materials have been shared with many investigators who have been actively engaged in hepatitis research."[73] The passage of clinical material between research laboratories can be interpreted—in a version of anthropological theories of gift exchange—as a means of incurring obligation, on the one hand, and securing a share of privileged access to knowledge, on the other. Almost certainly, such gifts help to cement bonds of loyalty, whether between patron and client or between equals.[74] Krugman at Willowbrook was mining a rich seam of hepatitis-infected blood from the mentally retarded children there.[75] Parceling out the serum to other hepatitis researchers over the years perhaps helped his survival in an increasingly hostile environment. It is also possible, of course, that his supporters were motivated simply by respect for Krugman's work, which many saw as valuable pioneering studies.

Conclusion

In 1951, the year the MRC report on the work of the Jaundice Committee was published, R. A. McCance, professor of experimental medicine at Cambridge, commented at a Royal Society of Medicine meeting: "The risk in any experiment depends very much on whether the investigator knows that he will always retain control of the situation. To inoculate somebody with icterogenic [jaundice-inducing] serum is a risk that I personally would never take, nor would I ever have cared to take it even before the risks were so well known, for once the inoculation had taken place I would have lost control."[76] MacCallum (key researcher in the British wartime jaundice transmission experiments) believed that the low mortality among troops who had caught hepatitis from contaminated yellow fever vaccine pointed to the probable containability of the infection. Others, in the spirit of McCance's statement, would say he was simply very lucky to have had no fatalities among his experimental subjects.[77] Some were quite ill, and the long-term effects, and transmission to contacts, are unknown and probably untraceable.

In terms of content, the wartime and postwar trials discussed above were fairly similar because they were seeking to establish routes of transmission and thus administered material from hepatitis patients—all sorts of bodily products—via alternative routes. The British MRC wartime trials included serum from patients who had attended venereal disease clinics, adding a risk of transmitting such diseases alongside hepatitis, although this was thought to be eliminated by the treatment of the serum before inoculation. Both sets of experiments had as their aim to improve understanding of the transmission of different types of hepatitis: in the wartime trials, in order to reduce outbreaks among troops (to which was added post-transfusion hepatitis); at Willowbrook, to combat the high prevalence of hepatitis among inmates. In the British trials discussed here, arthritis patients were encouraged to volunteer on the grounds of patriotism as well as the chance that a dose of jaundice could give them remission from arthritis symptoms. Parents of children in the Willowbrook trials, approached as the children were about to be admitted, were told that the studies would help all children in such institutions and that their child would be well looked after in the hepatitis clinic, where the course of a disease they might catch anyway would be closely monitored. So in both cases, consent was sought with a joint appeal to serving the common good and some possible individual benefit.

There are perhaps two chief differences, due to time and place, that might affect our retrospective judgement of these experiments. One is the state of knowledge. The MRC Jaundice Committee records and published material reveal a great deal more confusion over what types of jaundice or hepatitis were to be investigated than was the case when the Willowbrook experiments began over a decade later. The wartime experiments on both sides of the Atlantic clarified distinct routes of transmission, corresponding to two distinct types of hepatitis (others have since been discovered). It might seem, therefore, that Willowbrook replicated work already done. But to the investigators, the previous work was a useful but insufficient base for working out which types of hepatitis they confronted. The other chief difference is that between war and peace: in the view of the MRC Jaundice Committee, the needs of the war effort overrode a great many objections.

Of course, much criticism of a "medical ethics" nature has focused on issues of informed consent. The Krugman experiments, once "exposed," generated criticism because the subjects were minors and mentally retarded and thus doubly unable to give their consent. However, the mentally sound adults (arthritis patients) in the U.K. wartime trials were not fully informed of the nature of the experiments—especially the source of the infective material—and the archival evidence cited here shows that the Ministry of Health and the chair of the MRC Jaundice Committee were acutely aware of the potential objections. Their great uneasiness probably explains why the hepatitis transmission experiments ceased when the war ended and may well explain why publication of the full report was delayed.

This chapter has only looked in depth at the British wartime experiments, and a really full comparison with either the U.S. wartime or Willowbrook postwar experiments requires more digging in the archives. Enough has been revealed, though, to suggest that these hepatitis transmission experiments on human subjects were conducted out of a sense of necessity and with trepidation. Perhaps the chief difference in context between the wartime and postwar studies lay in the degree of secrecy, with increasing openness of debate in the postwar period eventually capsizing Krugman's work. Each generation, however, can only judge its own bioethical dilemmas: we should be wary of passing judgment on those of the past.

NOTES

A Wellcome Trust grant supported the research of which this essay is one outcome. I am grateful to the UK Medical Research Council (MRC) for permission to use their archives. Views and interpretations remain the author's responsibility.

The MRC archives here cited were still housed at the Medical Research Council headquarters in Park Crescent, London, at the time this research was undertaken; they were subsequently moved to the Public Records office at Kew.

1. II. K. Beecher, "Ethics and Clinical Research," *New England Journal of Medicine* 274 (1966): 1354–60.

2. W. Bynum, "Reflections on the History of Human Experimentation," in *The Use of Human Beings in Research with Special Reference to Clinical Trials*, ed. S. F. Spicker, I. Alon, A. de Vries, and H. T. Engelhardt (Dordrecht, 1988): 29–46, contrasts K. Mellanby, *Human Guinea Pigs* (London, 1945), which discussed without qualms Mellanby's wartime researches into scabies, using conscientious objectors as subjects, with M. H. Pappworth, *Human Guinea Pigs: Experimentation on Man* (London, 1967), which presented a highly critical view of a wide range of clinical experiments on human subjects.

3. Roger Cooter, "The Resistible Rise of Medical Ethics" (Essay Review), *Social History of Medicine* 8 (1995): 257–70.

4. Readers should note that the body of the paper is a shortened and revised version of chapter 2 of my thesis: J. Stanton, "Health Policy and Medical Research: Hepatitis B in the UK Since the 1940s," Ph.D. thesis, University of London, 1995.

5. E. H. Ackerknecht, "The Vagaries of the Notion of Epidemic Hepatitis or Infectious Jaundice," in *Medicine, Science and Culture*, ed. L. G. Stevenson and R. P. Multhauf (Baltimore, 1968): 3–16.

6. See S. S. Hughes, *The Virus: A History of the Concept* (London and New York, 1977).

7. G. M. Findlay and F. O. MacCallum, "Note on Acute Hepatitis and Yellow Fever Immunisation," *Transactions of the Royal Society of Tropical Medicine and Hygiene* 31 (1937): 297–308; G. M. Findlay, F. O. MacCallum, and F. Murgatroyd, "Observations Bearing on the Aetiology of Infectious Hepatitis (so-called epidemic catarrhal jaundice)," *Transactions of the Royal Society of Tropical Medicine and Hygiene* 32 (1939): 575–86.

8. F. O. MacCallum, interview by J. Stanton, 29 April 1992. Most of this section is based on this interview.

9. A. R. Hall and B. A. Bembridge, *Physic and Philanthropy: A History of the Wellcome Trust 1936–1986* (Cambridge, 1986).

10. W. H. Wright, *40 Years of Tropical Medicine Research: A History of the Gorgas Memorial Institute of Tropical and Preventive Medicine, Inc. and the Gorgas Memorial Laboratory* (Washington, DC, 1970).

11. Findlay and MacCallum (n. 7 above); MacCallum (n. 8 above).

12. Wright (n. 10 above).

13. Findlay and MacCallum (n. 7 above).

14. MacCallum (n. 8 above).

15. Ibid.

16. "Homologous Serum Jaundice," Memorandum prepared by medical officers of the Ministry of Health, *Lancet* 1943 (i): 83–8.

17. P. Curtin, *Death by Migration: Europe's Encounter with the Tropical World in the Nineteenth Century* (Cambridge, 1988) surveys the statistics but shows how public health measures reduced yellow fever mortality long before vaccine was available.

18. At this stage, the British were feverishly stepping up production of synthetic anti-malarials, having been caught in the same trap of reliance on German manufactures they had experienced in World War I, despite warnings in the interim. See Jennifer Beinart (Stanton), "The Inner World of Imperial Sickness: The MRC and Research in Tropical Medicine," in *Historical Perspectives on the Role of the MRC*, ed. Joan Austoker and Linda Bryder (Oxford, 1989), 109–35, esp. 122.

19. Editorial, "Jaundice Following Yellow Fever Vaccination," *Journal of the American Medical Association* 119 (1942): 1110.

20. This was Bedson, MacCallum's former boss, now serving on the Jaundice Committee, who was detailed to form the "Jaundice Research Team."

21. L. Witts to E. Mellanby, 4 Oct. 1943, in "Jaundice, Increase in the Incidence. Committee, Constitution & Members," folder 3217/1, MRC.

22. Hepatitis Sub-Committee, minutes of meeting at LSHTM, 20 Nov. 1942, in "Jaundice in Industry," folder 3217/4, MRC.

23. F. O. MacCallum, A. M. McFarlan, J. A. R. Miles, M. R. Pollock, and C. Wilson, *Infective Hepatitis: Studies in East Anglia During the Period 1943–47*, Medical Research Council Special Report Series No. 273 (London, 1951), 117.

24. MacCallum (n. 8 above).

25. Bradley cited (incompletely) G. F. Still, "On a Form of Chronic Joint Disease in Children," *Transactions of the Royal Medical-Chirurgical Society* 80 (1897): 52, where only passing mention is made of this effect; and the much fuller account in P. S. Hench, "Effect of Jaundice on Chronic Infectious (Atrophic) Arthritis and on Primary Fibrositis," *Archives of Internal Medicine* 61 (1938): 451–80.

26. MacCallum (n. 8 above). Military colleagues advised first investigating feces, as the most likely means of transmission of infectious hepatitis, but MacCallum prioritized the views of a Yorkshire GP who thought the disease might be carried by airborne particles; see W. N. Pickles, *Epidemiology in Country Practice* (Bristol, 1939). Fecal material was treated by centrifugation and ether extraction or freeze-drying, then disguised with vanilla or suspended in orangeade before use: MacCallum et al. (n. 23 above): 119.

27. F. O. MacCallum and J. D. Bauer, "Homologous Serum Jaundice: Transmission Experiments with Human Volunteers," *Lancet* 1944 (i): 622–27. Along with volunteers who had recovered from jaundice, "normal" volunteers were used; this

paper does not reveal their source. See also MacCallum et al. (n. 23 above): 127, for reference to pool of serum identified as source of jaundice; this Batch 034 was made from serum from 1,000 "supposedly normal" donors at blood banks.

28. L. Witts to E. Mellanby, 24 March 1944, in "Jaundice—Transmission to Volunteers," folder 3217/8, MRC.

29. A. Hood to E. Mellanby, 25 May 1944, folder 3217/8, MRC.

30. Dr. Methven, prison commissioner, to E. Mellanby, 10 May 1944, folder 3217/8, MRC.

31. L. Witts to A. Landsborough Thomson, 17 July 1944, folder 3217/8, MRC. By "support of the Council [the MRC]," Witts meant the support of Mellanby.

32. F. O. MacCallum and W. H. Bradley, "Transmission of Infective Hepatitis to Human Volunteers: Effect on Rheumatoid Arthritis" (Letter), *Lancet* 1944 (ii):228.

33. Minutes of fourth meeting, held at LSHTM, 11 July 1944, in "Jaundice Committee Minutes," folder MB39, MRC.

34. Dr. Janet Vaughan sat on the Transfusion Hepatitis subcommittee of the MRC Blood Transfusion Research Committee.

35. MacCallum and Bradley (n. 32 above).

36. MacCallum (n. 8 above).

37 L. Witts to E. Mellanby, 29 September 1944, folder 3217/8, MRC.

38. Witts to Mellanby (n. 37 above).

39. J. C. Knox to E. Mellanby, 11 February 1944, MRC 3217/8. This date suggests that the hunt for more arthritis patients coincided with the search for other options, outlined above.

40. E. Mellanby to J. C. Knox, 28 February 1944, MRC 3217/8.

41. MacCallum (n. 8 above). Note that he refers to animal experiments. In this recollection, nearly fifty years after the event, these were continuing alongside the human transmission experiments. This may indeed have been the case even though it is not evident from the published or archival sources.

42. MacCallum et al. (n. 23 above), 37.

43. F. O. MacCallum to E. Mellanby, 7 March 1945, folder 3217/8, MRC. Alice Stewart later established the link between radiation and childhood cancer. For a brief account of her wartime work in the context of her later career in epidemiology, see Gayle Greene, *The Woman Who Knew Too Much: Alice Stewart and the Secrets of Radiation* (Ann Arbor, 1999), 56–60. This mentions research on jaundice in munitions factories but not the transmission experiments discussed here.

44. MacCallum et al (n. 23 above): 134. In a large and busy Army clinic, syringes thrown into the autoclave and removed at random might not stay in for the requisite 10 minutes; Army informants confirmed the presence of minute quantities of blood in the needles and syringes: MacCallum (n. 8 above).

45. L. Witts to E. Mellanby, 7 March 1945, folder 3217/8, MRC.

46. Another possible outcome—a carrier state—was not recognized at the time, although subclinical infections (which may lead to a carrier state) were investigated through liver functions tests.

47. E. Mellanby to F. O. MacCallum, 7 March 1945, folder 3217/8, MRC.

48. Preface to MacCallum et al. (n. 23 above): iii. Authorship of the Preface is

not given. It is dated 4 September 1951 and states: "The investigation recorded here ended in 1947 and the report in its present form was accepted for publication not long afterwards." But it goes on to say that it was then postponed to make way for other reports and that it was published unrevised despite further knowledge on hepatitis accumulated meanwhile.

49. MacCallum et al. (n. 23 above), 143.

50. Ibid., 144.

51. Ibid., iii.

52. Editorial, "Homologous Serum Hepatitis," *Lancet* 1947 (ii): 691–92. See also report on section of general medicine, "Infective Hepatitis," *Lancet* 1944 (ii): 435–36.

53. MacCallum et al. (n. 23 above), 138.

54. Ibid., 37–45.

55. Ibid., 43.

56. Ibid., 32; children had been infected at school, but other family members could be infected through "relatively slight contact" with the sick child or through "close contact" with carriers.

57. There were also American wartime hepatitis experiments; the current study has not traced these.

58. Parental permission was crucial, but it was given on a general understanding of the nature of the experiments rather than detailed protocol. As with Mac-Callum's experiments, the exact nature of the material used for transmission trials was not spelled out to subjects.

59. For results of Willowbrook studies, see (inter alia): S. Krugman, J. P. Giles, and J. Hammond, "Infectious Hepatitis: Evidence for Two Distinctive Clinical, Epidemiological and Immunological Types of Infection," *Journal of the American Medical Association* 200 (1967): 365–73. For the view that these studies "represent an important contribution to our knowledge," see Editorial, "Is Serum Hepatitis only a Special Type of Infectious Hepatitis?" *Journal of the American Medical Association* 200 (1967): 407.

60. Beecher (n. 1 above); see also H. K. Beecher, *Experimentation in Man* (Springfield, IL, 1958).

61. W. Szmuness, H. J. Alter, and J. E. Maynard, eds., *Viral Hepatitis: An International Symposium* (Philadelphia, 1982).

62. R. W. McCollum, "Tribute to Saul Krugman, M.D.," in Szmuness et al. (n. 61 above), xxii.

63. Krugman et al.; and Editorial (both n. 59 above).

64. S. Goldby, "Experiments at the Willowbrook State School" (Corr.), *Lancet* 1971 (i): 749. For comment on Krugman's studies of the sort Goldby objected to, see Editorial, "Australia Antigen and Hepatitis," *Lancet* 1971 (i): 487–88.

65. Ushered in by an editorial comment immediately following Goldby's letter; see subsequent correspondence in same issue.

66. C. M. McKee, "Hepatitis B in Northern Ireland—Who Should be Immunised?" in submission toward part 2 of MFCM exam, 1988, chapter 3, "Historical Overview," 10.

67. S. Krugman, "Perspectives on Viral Hepatitis Infection: Past, Present and

Future," in *Viral Hepatitis: Etiology, Epidemiology, Pathogenesis and Prevention*, ed. G. Vyas, S. N. Cohen and R. Schmid (Philadelphia, 1978; Tunbridge Wells, 1979), 3–10.

68. William Muraskin, "The Silent Epidemic: The Social, Ethical and Medical Problems Surrounding the Fight Against Hepatitis B," *Journal of Social History* 22 (1988): 277–98, at 282; see also S. Krugman and J. P. Giles, "Viral Hepatitis: New Light on an Old Disease," *Journal of the American Medical Association* 212 (1970): 1019–29, for acknowledgment of a contract from the U.S. Army Medical Research and Development Command among others. Of course, the wartime jaundice investigations, both British and American, had been instigated because of military concerns as outlined above.

69. Krugman (n. 67 above), 6.

70. MacCallum et al. (n. 23 above), 128.

71. For history of antigen discovery, see: Stanton (n. 4 above), chap. 3; B. S. Blumberg, "The Australia Antigen Story" (Keynote Address), in *Hepatitis B: The Virus, the Disease, the Vaccine*, ed. Irving Millman, Toby K. Eisenstein, and B. S. Blumberg (New York, 1984), 5–31.

72. L. Goldman, "The Willowbrook Debate," *World Medicine* 7 (1971): 22.

73. Krugman (n. 67 above), 7.

74. See J. Stanton, "Blood Brotherhood: Techniques, Expertise and Sharing in Hepatitis B Research in the 1970s," in *Technologies of Modern Medicine*, ed. Ghislaine Lawrence (London, 1994), 120–33.

75. Krugman and Giles (n. 68 above), published in 1970, mention 2,500 serum specimens from 700 "patients" (children in the institution).

76. R. A. McCance, "The Practice of Experimental Medicine," *Proceedings of the Royal Society of Medicine* 44 (1951): 189.

77. Referring presumably to the American wartime experiments, Robert Purcell of the Hepatitis Viruses Section of the U.S. National Institutes of Health wrote parenthetically: "(several volunteers died of experimentally acquired hepatitis, and others became chronically infected)"; see Robert H. Purcell, "The Hepatitis Viruses: An Overview and Historical Perspective," in Szmuness et al. (n. 61 above), 3–12, at 4.

CHAPTER SIX

See an Atomic Blast and Spread the Word

Indoctrination at Ground Zero

Glenn Mitchell

On 12 May 2001, the American news broadcaster CNN posted a report on its Web site that contained admissions from the British Ministry of Defence (MoD) that it had used Australian servicemen in nuclear tests in Australia in the 1950s.[1] Australian newspapers and television programs carried the story in the following days.[2] The MoD strongly denied that it had used the soldiers as "guinea pigs" and repeated the claim given by the British government in 1997 to the European Court of Human Rights that it had never used humans in experiments in atomic weapons testing. However, the MoD admitted that New Zealand, British, and Australian soldiers were made to run and crawl through a contaminated area after atomic blasts at its Maralinga test site in the South Australian desert. "These were not nuclear tests as such, these were tests on clothing," the MoD said. "We were not testing people, we were testing the clothing. People have never been used as guinea pigs."[3]

CNN reported that the MoD described these soldiers as part of an Indoctrinee Force "whose role was to test equipment . . . which had been subjected to nuclear radiation."[4] What was this Indoctrinee Force? Was a multinational group of soldiers assembled solely to test the effects of radiation on clothing? If so, why the word *indoctrinee* in the name of their group? Did the British government use this force for other purposes, and if so what were they?

From October 1952 to October 1957, Britain's nuclear scientists conducted twelve nuclear tests in Australia. They exploded another twelve bombs in the Pacific Ocean on Malden and Christmas Islands between May 1957 and September 1958. Between 1953 and 1963 they conducted several hundred experiments known as the "Minor Trials." These tests, like the majority of the atomic explosions, were also held at Maralinga. In the eleven years of atomic tests, more than 35,000 British and Australian service personnel alongside 2,000 British civilians witnessed the twenty-four atomic blasts.

More than thirty years after the first blast, the Australian government appointed a Royal Commission to look into British nuclear tests in Australia.[5] The commission attracted wide media interest because its president, Mr. Justice J. (Diamond Jim) McClelland, a staunch republican and former Federal minister, relished the opportunity to question British civil servants. He accused the British government of failing to disclose information about its tests and of "dragging its feet" in providing information and harangued past Australian governments, especially that of Sir Robert Menzies, for being a "lickspittle of the British." He was dogged in his pursuit of evidence. When William Penney (later Lord), the director of the British atomic tests claimed he was too ill to travel to Australia to give his testimony, McClelland flew to England and convened a sitting in Penney's residence. Media coverage of the 118 sitting days and the commission's reports revealed a story hitherto kept secret from most Australians. Without this Royal Commission, it is doubtful that previously classified documents would have become public.

This chapter examines one aspect of the many experiments carried out at Maralinga and attempts to place the experiment in the context of the burgeoning human experimentation for medical research during the cold war.[6] The opportunity to make and witness explosions "brighter than a thousand suns" proved irresistible to American, Australian, and British military and scientific authorities.[7] At Maralinga, British and Australian governments brought together a select group of men, largely senior officers, known as the Indoctrinee Force, or I-Force (IF). Accounts of its work are in a history of British atomic tests in Australia commissioned by the Federal Department of Resources and Energy, the final *Reports of the Royal*

Commission, and references in popular accounts of the British atomic tests by journalists.[8] These largely narrative accounts provide little analysis of the IF's work or the reasons for its formation. The original reports, however, provide much valuable detail.

In its submission of documents to the Royal Commission, the British government provided several reports on the IF prepared by the Atomic Weapons Research Establishment (AWRE) at Aldermaston, England, at the time of the Australian tests.[9] Moreover, during a successful claim for compensation by a former Australian serviceman who had worked at Maralinga in 1956, these same reports and others came to light. This chapter draws on this archival material.

Despite the AWRE reports and a heated debate by atomic veterans and their supporters that has brought forward new evidence such as the MoD's admission on the "clothing tests,"[10] what the English historian R. H. Tawney called the "shrivelled tissue," remains thin.[11] For example, Penney burned his papers in a backyard blaze prior to his death, and the most extensive collection of material on his work, his personal papers at the Imperial College Archives, London, throws little light on his work in Australia.

There is much that we do not know about the Indoctrinee Force. We do not know what each man thought of his work—there is no history of the IF, biographies or autobiographies of its members are yet to be written, and there is no published list of those who made up the IF. Even some members of the IF are unable to explain why they were sent to see an atomic bomb explode.[12] We do not know the health status of each member of this group—there is no health study of the IF or individual members. While declassified documents tell something of what the IF's military superiors were thinking when they formed the IF and sent it to watch atomic bombs explode in the hot Australian desert,[13] we do not know the outcomes of the work of the IF since no Australian or British agency has published any results of the IF's work.

However, this much is clear. The officers who made up the IF were part of a three-stage experimental training exercise. In the first part, IF members would receive lectures and literature on atomic physics and the explosion of atomic bombs. The second part would see the IF travel to Maralinga and experience an atomic blast at close range. To complete the exercise as the third stage, the IF members would go back to their units to spread the word about their positive reactions to atomic blasts. This exercise was about indoctrination—both in the classroom and by firsthand experience. In broad terms, those in charge of the IF wanted to "indoctrinate" its members with the "intimacy" of atomic explosions, to use personal experiences to allay the fears of their men, and to spread the word about the "safety" and operational "utility" of working and surviving at ground zero.

This chapter examines how two governments used many people as objects of indoctrination. This was an experiment, albeit an unusual one, conducted in complete secrecy. Little was known of the work of the IF until the Royal Commission in 1985. The formation of a force of senior military officers with the objective of acquiring information about atomic blasts so that its members could pass on positive accounts demands a different way of thinking about human research subjects and human experimentation. While the terms "experiment" and "guinea pig" seem strange in conjunction with the Indoctrinee Force, I argue that these men were used as research subjects. Those responsible for the IF asserted that its work was in effect done on behalf of British and Australian people. I also argue that the indoctrination of these men was part of a larger conditioning and education process—that of a widespread public acceptance of atomic weapons. The chapter begins with the background and history of British nuclear tests in Australia. It then examines the formation and work of the IF. The final section challenges the conventional explanations of the consequences of the work of the IF.

The Bombers Prepare

The dropping of atomic bombs on the Japanese cities of Hiroshima and Nagasaki continues to raise ethical, legal, medical, moral, and political debate.[14] Writers such as P. M. S. Blackett and M. J. Sherwin have argued that the potential for unprecedented destruction was sufficient to guarantee peace.[15] The historian Susan Lindee has argued that the bomb was "a frightening manifestation of technological evil, so terrible that it needed to be reformed, transformed, managed, or turned into the vehicle of a promising future. It was necessary, somehow, to redeem the bomb."[16] Australian politicians and atomic scientists such as Ernest (later Sir Ernest) Titterton and Philip Baxter (later Sir Philip) ascribed a broad economic vision to the raw material of uranium and the production of atomic energy.[17]

In the 1950s, the governments of the United States and the Soviet Union embarked on ambitious programs of atomic tests, choosing areas within their borders for their tests. For British authorities, however, the development and testing of nuclear weapons posed significant problems. When the U.S. Congress passed the Atomic Energy Act —the McMahon Act—in July 1946, it prohibited American scientists from exchanging any information from its nuclear research program with any nation.[18] The legislation made no concessions to the past contributions of scientists from outside the United States to atomic science. For example, prominent British atomic scientists such as Penney and Titterton had worked on the Manhattan Project. Penney had won over General Groves, "the formidable and somewhat Anglophobic chief of the Manhattan Project"[19] and was widely admired at Los Alamos. Yet the McMahon Act gave his work and scientific standing and that of other British scientists few concessions. The sharing or exchanging of information between former American and British colleagues did not work as it had during the war.[20]

The British government realized that if it wanted to develop either an atomic bomb for military purposes or nuclear energy for peaceful purposes, it would need to use its own resources. Put simply, the British had to "get it on their own hook."[21] Margaret Gowing, the official historian of Britain's atomic energy project, lists four possible reasons for Britain's atomic project: to advance research and development not only in all the sciences of atomic energy but also for its practical application; to build atomic weapons; to use atomic energy to generate electricity; and to use radioactive isotopes for medicine and industry.[22] National pride and perceived security reasons could be added to this list—Britain would show the rest of the world that it could become a nuclear power and saw it as a means to security. With divisions between communist and noncommunist nations becoming more entrenched in the late 1940s and 1950s, British authorities believed that a nuclear capacity would be important in a time of war.[23]

For development of atomic weapons authorities needed a testing site. The British government rejected testing within the United Kingdom, with Penney in particular not wanting to see fallout polluting local waterways.[24] The British attempted to restore the relationship that the McMahon Act had severed, yet despite tripartite talks (America, Britain, and Canada) and bilateral agreements between the British and the Americans, these did not include a test site.[25] Britain approached its Commonwealth partner Australia, who subsequently gave permission initially to explode atomic bombs at the Monte Bello Islands off the Western Australia coast in 1952 and 1956, and then at Woomera and Maralinga in South Australia in 1953 and 1956–57 respectively.[26] The British began its series of atomic trials in Australia with the explosion of its first device on 3 October 1952 and continued until 1963. The British also conducted hundreds of so-called minor trials. These were tests with components of weapons, which were called Kittens, Tims, Rats, and Vixen A and Vixen B. Britain later admitted that it hid the real purpose of these trials. Final tests in the "minor trials" in the early 1960s were aimed at investigating the dispersal of radioactive material.[27]

The Bombers Arrive

On 14 August 1956, William Penney, who supervised the British tests, and Mr. Howard Beale, the Federal Minister for Supply, held a press conference to announce details of four atomic explosions to take place at Maralinga. Beale later told Federal Parliament that this was "a very valuable press interview . . . the biggest press interview ever held in Australia," which was "part of policy of giving the Australian people the fullest possible information about this matter, to which they are clearly entitled . . . [and] was organized to present purely factual information about these tests."[28] This press conference was in stark contrast to the Monte Bello tests, where little information was made available to the Australian public prior to the tests and even British politicians were not fully apprised of the tests.[29]

Penney talked at length about the tests, giving careful and specific emphasis to their safety precautions. He went to considerable lengths to assure Australians that the tests would be safe and that they posed no risks to military or civilian personnel. Safety was of "outstanding importance," "meticulous care" would be taken, and a safety committee of distinguished Australian scientists would examine and approve the proposed explosions in the coming months.[30]

Two days earlier, the *Sydney Morning Herald* had published Howard Beale's explanation of the atomic tests. Like Penney, he paid considerable attention to the safety aspects of the tests, arguing that the Australian government had given its consent only "upon conditions of complete safety." The British government and its scientific advisers had given assurances that the tests would be "harmless to Australian life and property." He also wrote of "rigid safety requirements," "quite harmless" increases in radiation, and "authoritative assurances of safety" from the British Medical Research Council and the American National Academy of Sciences. His final words were: "We will continue to assist Britain (and ourselves) by the careful carrying on of atomic tests at Maralinga under conditions which will ensure complete safety, and will enable us to expand our atomic knowledge in the service of mankind." If Beale had any doubts about the safety of the tests—he used the phrase "complete safety" on four occasions—this testimony to the bombers' commitment to human and environmental protection gave no indication of it.[31]

Because these tests had military objectives, neither Beale nor Penney told the press conference everything that would take place at Maralinga. For example, there was no reference to the formation or deployment of the Indoctrinee Force. The emphasis was on safety, care, and the absence of risk to witnesses at the site and to the Australian public. There was good reason for this approach. The memory of Hiroshima and Nagasaki was still fresh in the memory of Australians, and the Monte Bello tests in 1952 and 1956 had not allayed fears that atomic tests were not without risks. The local press had not provided authorities with the glowing testimonies they had hoped for, and after the G2 explosion on 19 June 1956, newspapers had headlines such as "Radioactive rain reported in North" and "A-dust danger is here." At a demonstration in Perth in July 1956, protesters against the Monte Bello tests waved banners about the use of Australians as guinea pigs.[32] A civil servant from the British Ministry of Supply wrote back to Whitehall: "It is undoubtedly very difficult for people in the United Kingdom to appreciate how powerful public opinion in Australia is in connection with atomic trials, but it is a fact that on occasions connected with the Monte Bello tests the pressure on the Government from the press and the Opposition was so great that there was real danger they would be obliged to refuse any further tests in Australia."[33]

For each objection, however, there was the same response—eminent scientists assured the Australian government that all care would be taken, and reports from prominent medical and scientific authorities supported their assurances. For example, when Beale was asked in the Federal Parliament about the concerns raised by Dr. John Wolfe, an American atomic energy scientist, about potential hazards from radioactive fallout from the tests, Beale admitted he had not seen the report. "I do not think I have seen the statement attributed to Professor Wolfe," he said. "There are odd professors and scientific men in various parts of the world who hold views of that kind. I am bound to tell the honourable member, however, that the overwhelming body of disinterested and responsible opinion is very reassuring on the safety aspects of the atomic tests. In any event, the honourable member may be assured that the Australian Government, advised as it is by very eminent scientists, will make sure that there is no danger to Australians."[34]

Beale's interpretations of the risks and his construction of safety at Maralinga echoed public commentary in England and the United States that nuclear tests were safe. He was wrong, however, to see those scientists who supported the tests as purveyors of "disinterested and responsible opinion." The scientists to whom he was referring, including those in the Atomic Energy Commission and the AWRE, were hardly disinterested—they were partisan, and with few exceptions, highly committed to the development of nuclear weapons.[35] Nor did Beale understand the difficulties in extrapolating the effects of high doses of radiation, such as those recorded at Hiroshima and Nagasaki, to the low levels of radiation that Penney and others asserted would occur at Maralinga. It is also unlikely that American or British researchers would have briefed Beale on effects of low-dose whole-body radiation. Moreover, as we shall see later, there is evidence to indicate that "disinterested" scientists and British civil servants had little difficulty in convincing Beale about assertions of safety.

The Indoctrinee Force

In August 1984, when a Royal Commission, the highest form of public inquiry available to an Australian government, commenced its investigations into the British nuclear tests in Australia between 1952 and 1963, it could barely speculate as to the extent and complexity of evidence its work would reveal. Its report eventually provided a different picture of the detonation of atomic weapons than the conventional wisdom constructed by British and Australian authorities. It cast doubt on, openly challenged, and in some cases rejected earlier assertions of safety, risk, and appropriate precautions during the tests. And in doing so, the report revealed more to Australians than they had hitherto known about the British nuclear tests. Issues not debated in Australia at the time began to receive considerable, albeit belated, attention.[36]

One such issue that had received almost no publicity during or after the tests was the work of the Indoctrinee Force.[37] Witnesses and hitherto secret documents raised many questions about the IF: Why was it formed? What evidence did the British and Australian governments seek from the

IF? What results did the IF return to its superiors? And what results did its commanders give the IF? How did the government believe the IF would benefit its military objectives? Did the IF's work have medical and/or health implications? Was this a human experiment?

The AWRE at Aldermaston[38] set out the tasks of the IF as follows:

- Two hundred and fifty serving officers from the armed forces of the United Kingdom, Australia, and New Zealand will be exposed, at a safe distance, to the flash, heat and blast effects of an atomic explosion.
- A conducted tour of the firing area will be made before and after the event to enable indoctrinees to observe and appreciate the effects of the explosion on the ground and on items of Service equipment, vehicles, and structures exposed for the Target Response Trials.[39]

The IF eventually consisted of 283 men; 178 from the United Kingdom (172 officers and 6 civilians); 100 from Australia (74 officers, WO and CPO, 25 from other ranks, and 1 civilian); and 5 officers from New Zealand. The U.K. representatives received lectures in London and Singapore on the science of atomic bombs and detonation, while the remaining IF members received similar lectures and instruction at Maralinga.[40]

If authorities had in mind the collection of information about possible connections between atomic explosions and human health, they did not articulate this objective in planning, operational, or evaluation reports on the IF. The effects of atomic explosions on human and physical environments were well known before the British exploded its bombs at Maralinga. The dropping of bombs on Hiroshima and Nagasaki had taken scientists and medical investigators from the laboratory and animal studies to the field, where the effects of atomic explosions on human health could be clearly documented. Apart from one work, *The Effects of Atomic Weapons*, prepared by the Los Alamos Scientific Laboratory for the U.S. Department of Defense and the U.S. Atomic Energy Commission, the considerable body of work on post-atomic health research does not appear to have been used in the preparation of the IF. Force commanders provided a small library at Maralinga as a source of "study for the serious minded." It listed no works on ethical issues or health studies.[41]

Moreover, Penney was no novice in this area, having worked from 1944 to 1945 as the principal scientific officer for the British Department of Scientific and Industrial Research at Los Alamos, New Mexico. While American authorities thwarted his ambition to witness the first bomb on Hiroshima, he was an observer on the plane that dropped the second bomb on Nagasaki. He was also one of the first Western observers to tour the Japanese bomb sites.

If members of the IF were to provide their military superiors with specific health and/or medical information, there is no record of the collection of this data.[42] Members of the IF were not given a medical test before the blasts, nor were they tested or examined after the tests. IF members were issued with film badges at Maralinga that detected changes in radiation—some IF members testified that their badges were never examined, while others only found out for the first time at the Royal Commission's hearings about the readings of the Maralinga badges.[43] There is no record of the collection of data about psychological reactions either. Members of the IF were not given a psychological profile before, during, or after the atomic tests. Moreover, there appears to have been no systematic gathering of any data from members of the IF.

So if no detailed tests of a medical, health, or psychological nature were carried out on members of the IF, what purposes and objectives did the AWRE and the British and Australian military authorities ascribe to these men? Was the purpose nothing more than the objectives spelled out by E. R. Drake Seager, the IF coordinator at Maralinga? He wrote: "The primary purpose of the exercise was to get Service personnel accustomed to what nuclear war was all about. . . . In addition to familiarising people with the effects of a nuclear explosion, the aim was to 'spread the word' and the IF was chosen from middle-ranking officers who were seen to have good career prospects and who could and would be expected to return to their units to lecture from first-hand knowledge on the effects of the test (or tests) they had witnessed."[44]

He made no attempt to say what was "the word" they were meant to spread. However, I think it is fair to assume that "the word" was a positive assessment of being close to an atomic blast. He concluded:

> [The] IF project was completely successful. It established that it was feasible; it was possible to fit a large number of people into a nuclear test without too much disruption. "Indoctrination" (in the proper sense of the word) was achieved; force members had understood what they had been told and seen and were thus in a position to spread the word. There were unexpected bonuses through the delay in that an expanded programme of lectures, with more experts, could be given and IF members helped to set out various Tar-

get Response items; thus they had a direct involvement in targets with which they had practical experience. The Target Response Group itself benefitted by being able to perform more detailed experiments thanks to the IF help.[45]

There is no doubt that the atomic tests at Maralinga were experiments and that the members of the IF and others were experimental objects. Penney, Beale, Titterton, and others used many words to describe the series of atomic explosions: "trials," "tests," and "scientific experiments."[46] And there is no doubt that the AWRE and the Australian Atomic Weapons Test Safety Committee (AAWTSC) saw the IF as a valuable part of the experiments carried out at Maralinga. The men of the IF were to provide information to the AWRE and the Army Operational Research Group (AORG) on five critical areas: attitudes toward nuclear weapons, issues relating to morale, the accumulation of detailed personal knowledge about atomic blasts for the purpose of instruction, the construction of further evidence to support the assertion that these tests were "safe," and the strengthening of the concept of civil defense.

Human Guinea Pigs and Experimental Subjects?

Each series of British atomic tests had code names. The first four tests at Maralinga in 1956 had the code name of Operation Buffalo; the three tests in 1957 were known as Operation Antler; and the so-called minor trials were code-named Kittens, Tims, Rats, and Vixen. These trials attracted considerable attention at hearings of the Royal Commission in 1984. For example, the commission heard and dismissed allegations that the Buffalo trials used "mentally defective people" in atomic experiments.[47]

On the question of "human guinea pigs," the Royal Commission was less emphatic. It noted Drake Seager's assessment in 1956 that this was not the case and heard him restate this observation in his evidence to the Royal Commission: "I repeat that the members of the IF were not being used as human 'Guinea Pigs' to test the effects of a nuclear explosion; all were positioned at distances and in locations calculated to be safe." Another witness, A. C. Flannery from the Commonwealth government's Department of Supply, supported Drake Seager's testimony: "The indoctrinees were there for observation purposes only. They were very senior officers and I cannot see that they would be used as guinea pigs."[48]

This is not a view without challenge. In *The Ethics and Politics of Human*

Experimentation, Peter McNeill examines many cases of human experiments. He includes Maralinga, not because the tests "were designed as experiments on human subjects as such," but "because of the obvious risk of harm to humans and the commonality of many of the ethical issues in nuclear testing and experimentation on human subjects."[49] Recent evidence from U.S. congressional hearings support this argument.[50] Moreover, some members of the IF testified that the use of words such as "calculated to be safe" is problematic and ignores questions such as: What standards and measurements were used to calculate a distance to safety ratio? And how were these standards and measurements arrived at?

Robert Walgate, drawing on the work of the prominent radiation physicist Joseph Rotblat, argued that standards of safety for ionizing radiation were in "a state of flux" in the 1950s.[51] Walgate sketches the following history: The International Commission on Radiological Protection (ICRP)[52] revised its recommendations and standards for "tolerance" levels and "threshold standards" on three occasions between 1951 and 1958. Of interest is its recommendation in 1954, which said in part that "no radiation level higher than the natural background can be regarded as absolutely 'safe.'" The ICRP also said of "temporary" exposure: "Since it is generally impossible to predict how long a person may be occupationally exposed to radiation . . . it is prudent to assume that it may continue throughout his life. Therefore 'temporary' exposure at levels higher than the permissible weekly dose should not be permitted." There is no evidence to show that the British or Australian authorities took these recommendations into account for members of the IF.[53]

It seems that those in charge of the IF paid scant heed to the question of radiation risk. There was little if any monitoring of radiation levels at Maralinga of IF members and no follow-up studies.[54] Other projects such as the Manhattan Project in the 1940s and the American Bravo tests at Bikini Atoll in 1954 that involved exposure to radiation also included attention to safety levels and efforts to monitor those exposed during the tests and afterwards.[55] This was not the case for the IF. It is not clear why this was not done at Maralinga but was done elsewhere.

Attitudes Toward Nuclear Weapons

Mr. L. J. Holman, a member of the Army Operational Research Group within the Department of the Scientific Adviser to the Army Council wrote a report on the work of the IF in which he noted that the "the object of their 'indoctrination' was to give them a better idea of the nature and possibilities of nuclear warfare than could be inculcated by lectures and 'cold' or faked demonstrations alone."[56] This report and others on the men of the IF is unclear as to the meanings of critical words such as "indoctrination" and "better." It is important to note that at no stage in either planning, operational, or evaluation reports is the word "indoctrination" given a precise definition. We are left with the impression that this process was ad hoc, based naively on the idea that if the men actually witnessed an atomic blast, they would learn *something;* the blast in itself would have an inherent, if undefined, utility.[57]

Holman's report not only noted the uncertainty in the way indoctrination might work. It also bedeviled the measurement of attitudinal change in the IF toward nuclear weapons. He wrote:

> In order to ascertain whether any measurable change had occurred in the attitudes of the Indoctrinees towards nuclear weapons as a result of their experiences, it was suggested to AORG that pre-bomb and post-bomb questionnaires be produced by them and administered with this end in view. Between arriving at the camp and completing their post-bomb questionnaires, the Indoctrinees had had an intensive course of scientific lectures on the subject of nuclear warfare and had assisted in preparing the range: so that any changes in their knowledge, opinions and attitudes cannot necessarily be ascribed to the witnessing of the explosion itself or to the examination of the target response area.[58]

However, despite the lecture material, the AORG concluded that the IF had a "definite increase in knowledge . . . after the explosion. They also appeared to be more ready to accept a nuclear missile as a tactical weapon after the explosion than before, and the indoctrinees before the explosion were more ready to accept one than were the control group."[59]

Issues Relating to Morale

The AORG concluded further that the formation of the IF and its exposure to an atomic blast was a success: nearly all the indoctrinees said they had gained "something from their experience," with approximately half of them saying they "would now be able to lecture and train more convincingly, with more authority, on the subject." The results, despite their internal inconsistencies, impressed Holman, who recommended to the AORG that if AWRE wanted to "assess the probable morale and shock effects of a nuclear explosion, formed units should be used in a realistic exercise."[60]

Holman's introduction of "morale" gives an interesting insight into one of the critical, yet largely unspecified, objectives of the IF initiative. It is clear from Holman's report that while issues of morale were not identified as principal objectives of the IF's work, morale building was in fact an important area that the MoD and the AWRE wanted to develop. Put simply, those in charge of the IF wanted its members to see an atomic bomb explode, and as a consequence of this experience and the lectures provided as part of the exercise, take a positive report back to their men—Drake Seager's reference to "spreading the word" can be read as "take your boost in morale from the bomb site back to your men and tell them about how it made you feel, how Britain has succeeded in developing a powerful strike weapon."

In another report, *Reaction of IF Group to the Nuclear Explosion*, IF members observed that the "morale effect of the weapon is extremely difficult to assess accurately from this type of indoctrination." Their observations on the flash, heat, and noise of the blast indicated that the IF was clearly impressed by the blast: "The heat effect experienced was remarkable. . . . It was as if someone had placed a very hot iron or an electric fire unpleasantly close to one's neck. It appeared to last about two seconds and one could readily appreciate the tremendous casualty producing effect of thermal radiation at close ranges."[61]

The report by the commander of the Indoctrinee Force was more glowing than Holman's or the IF members. He saw the work of the IF as a resounding success. However, rather than forming another IF as recommended by Holman to replicate the IF's work in further tests, he preferred to see "a complete unit such as a Parachute Battalion, or even a complete unit of all arms with the normal command structure, put through the mill properly and the lessons learned assessed by a small party of high grade observers."[62] Various annexures to this report reinforced his enthusiasm.

One IF member, Colonel G. W. H. Peters, wrote "The best way to train an individual is by indoctrination," and he noted the "staggering effect on morale of a nuclear explosion."[63] While Peters did not elaborate on what he meant by "staggering effect," it is clear that he was referring to the positive contribution that the blast had made to boost morale. Another IF member, Colonel W. A. R. Ross, was also interested in issues of morale: "It is further suggested that indoctrinees of all arms should receive instruction from an RAMC officer on the medical aspects of nuclear warfare, including casualty estimation and psychological effects."[64] Lt. Colonel R. K. Gregory, also an IF member, was also impressed by the morale boosting powers of witnessing an atomic blast and wrote that the "maintenance of morale is more important than ever."[65] Brigadier L. H. Gordon, also an IF member, recommended the development of a "trailer mounted decontamination unit consisting of a steam jenny, a high pressure wash and an industrial type vacuum cleaning unit," prototypes of which could be tested at future atomic trials. Gordon also recommended changes to training for future indoctrinees, including training in the handling of contaminated material. He noted that the "psychological fear of radioactivity must be overcome and confidence established that it can be handled in safety provided the right methods are used."[66]

The Construction of Safety

Scientists and politicians worked very hard before and during the tests to put the best possible spin on safety and the elimination of risks. When we look at who was asserting the safety of the tests, we see powerful figures who acted as commentators on scientific matters. They also acted with strong political motives within a social context that encouraged their science to be political. Earlier work on radiation and scientific controversies has examined the role of political interests.[67] We now know that many of these efforts were careful constructions of events with careful wording and in some cases either the sanitizing, removal, or both of various pieces of information. Three sources support this conclusion: Penney's assertions at his press conference in 1956 and his testimony to the Royal Commission in 1985; Titterton's public assertions on safety at the time of the tests and his private concerns; and the construction of Beale's press statements before and during the British tests.

At his Sydney press conference, Penney said that the tests would be carried out in complete safety and with a total absence of risk. At the Royal Commission, however, he admitted that there were risks and that authorities at Maralinga had not told all participants at the tests, including members of the IF, about all the risks. Consider the following exchange under cross-examination during the commission:

Q. You were responsible, you told us, and you told us the way you were responsible, for the decision with respect to the regulations for the level of exposure of personnel?

A. Yes.

Q. In exposing someone in that way, you knew that there was some attendant risk?

A. Yes.

Q. You were so advised?

A. Yes.

Q. Did you take all the steps to tell those who might be exposed, what the risks were?

A. No. The person I would regard as responsible for doing that would have been the safety officer, the range officer.

Q. Do you know whether or not it was ever explained to people who were involved that in being exposed to levels of radiation there was an attendant risk?

A. No. I do not. What I believed happened was the radiological protection people there proscribed people moving into the area, and they monitored them, but whether they went and told every man, "This is a dangerous area," etc., I do not know. I would regard that as part of their duties.[68]

Safety, indeed complete safety, became strongly associated with the Maralinga atomic tests just as this assertion had characterized American nuclear tests. In 1955 the U.S. Department of State sent a telegram to the United Kingdom saying in part, "Nuclear weapons central part of defensive capability. . . . Our own studies have demonstrated that no significant health hazard results from nuclear test explosions. We are presently furthering such studies to provide additional information this matter."[69] Professor Ernest Titterton, a member of the Australian Safety Committee, which approved the firing of the atomic devices, asserted that the atomic blasts would provide no risk to the public.[70]

During and after the tests, the claim of safety was repeated many times. The editorial of the *Sydney Morning Herald* claimed that the fifteen postponements of the first test added to the safety factor: "No breach of the strict code of precautions was allowed, no matter how great the inconvenience and loss. In these matters the basic assurance comes from trusted scientists like Sir William Penney."[71] Professor L. H. Martin, chairman of the Australian Safety Committee, told Minister Beale: "You can give the public an unequivocal assurance that there is a complete bill of health; all dangerous fall-out has been deposited and the remaining fall-out is completely innocous."[72] Scientists and ministers made similar claims following the other three atomic tests.

While Titterton publicly asserted the complete absence of risk or danger to all Australians from the nuclear tests and gave generous meanings to the concept of safety at Maralinga, his private assessment was somewhat more cautious. In a letter to Howard Beale on 14 October 1957, he said: "I am pleased that we have managed to get through this series of tests without any reports of radioactivity detected in continental Australia and hope that we may be able to maintain this record in future operations, but, as you know, we cannot guarantee this."[73]

His private evidence throws his public confidence into sharper relief. For example, scientists kept adding various soil and air tests as the bomb trials went along. Strontium measurements were not taken at the Maralinga tests. However, in October 1956, for a man who was given to making frequent public pronouncements about the safety of the tests, Titterton appeared to have some reservations. He was clearly unsure about both the levels of fallout and their possible impacts when he wrote to the Atomic Energy Research Establishment at Harwell, Berkshire, saying, in part, "I think now that it is of growing importance that we should make some strontium measurements."[74]

Titterton's public confidence and assertions about absolute safety were not always consistent. In his book *Facing the Atomic Future*, he said of atomic tests: "There is an element of risk in carrying out such tests"; and "in this imperfect world safety can never be guaranteed . . . we have to balance the production and testing of weapons, which do so much to preserve our present uneasy peace."[75]

In addition to the scientific and technical significance of the atomic tests, Titterton and others were also concerned about the construction of information so as to continue the illusion of absolute safety. While Beale may have told the Federal Parliament about the efficacy of "disinterested and responsible opinion" from "eminent scientists," two things are clear: Titterton and others were *not* disinterested, and he and others used their scientific standing to dominate and construct pictures of the tests that fitted their political views. Titterton rejected dissent, especially of his views, and relied on the status of "eminent" scientists to reinforce his own views. When Titterton wrote to AWRE about analyses of soil samples, he said,

"I believe that a small but continuing systematic soil and animal bone survey here in Australia is of great importance. A matter like this can have considerable *political significance* and it is important to have as good measurements as we can get."[76] In another letter to AWRE, Titterton commented on adverse criticism of the work of the AWTSC and under-took to remove this perception: "[I understand] a suggestion has been made that the Safety Committee has been given too little information for it to properly assess the safety position. I don't know where that one came from but I will knock it firmly on the head tomorrow."[77]

Titterton had a long association with the national radio broadcaster, the Australian Broadcasting Commission (ABC). He had given two talks in 1952 on the "Ethical Dilemma of Science" and "Education for the Atomic Age from a Psychological Point of View," and in 1956 he wrote to the ABC suggesting another series of talks to clear up misconceptions about the atomic tests. Once again his focus was on the political advantages of his talks. He left the ABC in no doubt that his views were the ones that the public needed to hear. He wrote: "There is a good deal which can be said on this subject particularly as political pressure and ill-informed adverse comment has been made against the atomic weapon tests being carried out in Australia. The whole problem could do with a thorough discussion for the benefit of those people who really want to have the facts."[78] Not sur-prisingly, a prominent nuclear researcher has concluded that his 1956 book, *Facing the Atomic Future*, and other popular writings "are frankly nu-clear propaganda."[79]

Howard Beale also gave many public assurances about safety—state-ments at the joint press conference and in Federal Parliament—about the complete safety and absence of risk at the tests. Yet on occasion, his words were sanitized so as to support such assertions. On 24 August 1956, Beale released a press statement on the use of small animals in the Maralinga atomic blasts. The statement said in part, "The purpose of these investi-gations is to increase our knowledge of how to protect the civilian popu-lation in the event of an enemy atomic attack."[80] However, these were not his original words. British authorities had edited Beale's statement. The original draft of Beale's statement had the words "exposed herbage." The Office of the High Commissioner for the United Kingdom in Canberra was very concerned about these two words and wrote to Sir John Bunting in the Prime Minister's Department. The note, dated 24 August 1956 and headed *Top Secret: Immediate*, said the British government wanted these words deleted "as this might be taken to imply widespread contamination in the area outside the test." The note also deleted "radiation" and replaced it with "atomic attack."[81]

This was not the first time that British authorities had revised Australian press releases so as to give a picture of "complete safety." When the matter of live animals was first raised, the original draft had the sentence "The possible effects of the ingestion of radio-active fallout (by man and animals) will be among the subjects studied." On 25 January 1956, A. G. R. Rouse of the High Commission of the United Kingdom wrote to the Prime Minister's Department suggesting a number of amendments to the draft document. He said, "The omission of '(by man and animals)' in the first paragraph is suggested as it might lead to imply that direct experiments on human beings will be carried out."[82] The Australian authorities amended the document without question.

The Bombers Assessed

Five sources of information challenge and dismiss many of the assertions made by Penney, Titterton, and Beale. These sources, either considered separately or together, provide a different picture from that constructed by the Maralinga bombers. The first source is strong circumstantial evidence.

According to Hazel O'Leary, Secretary of the U.S. Department of Energy, the U.S. government conducted more than four hundred radiation experiments during the cold war, with more than 16,000 people being used as guinea pigs in these experiments.[83] With the recent release by the Department of Energy of extensive documentation about these human radiation experiments, we know also that the U.S. Army used soldiers as human guinea pigs during atomic tests.[84] Throughout the final report of the Advisory Committee on Human Radiation Experiments that researched and prepared the release of the many documents, the term "guinea pig" is used and accepted without qualification.

It is also clear what the U.S. Army did and why it used its soldiers as guinea pigs. The army deployed troops at bomb test sites to indoctrinate them about atomic warfare and to later carry out biomedical tests.[85] Dr. Richard Meiling, the chairman of the Armed Forces Medical Policy Council,[86] in a memorandum to the Department of Defense argued that "troops should be placed at bomb tests not so much to examine risk as to demon-

strate relative safety." Meiling wrote: "Fear of radiation is almost universal among the uninitiated and unless it is overcome in the military forces
it could present a most serious problem if atomic weapons are used. . . . [It]
has been proven repeatedly that persistent ionizing radiation following air
bursts does not occur, hence the fear that it presents a dangerous hazard
to personnel is groundless. . . . Positive action [should] be taken at the earliest opportunity to demonstrate this fact in a practical manner.[87]

Meiling recommended the deployment of troops near ground zero
rather than their movement into the blast zone. This exercise he believed
"would clearly demonstrate that persistent ionizing radiation following an
air burst atomic explosion presents no hazards to personnel and would effectively dispel a fear that it is dangerous and demoralizing but entirely
groundless."[88] There is no evidence that directly links Meiling's proposals to Penney and his Maralinga bombers. However, it is clear that the IF
was used to test and boost morale, to indoctrinate troops so as to cure them
of their "mystical fear of radiation,"[89] and that its purpose was very similar to the proposal Meiling had for U.S. troops.

The second source of information is contained in the evidence produced in a joint dissenting opinion of Judges De Meyer, Valticos, and
Morenilla of the European Court of Human Rights in the case of two
British nuclear veterans. The Court examined the claims of two British
service personnel who were present at Britain's atomic weapons trials on
Christmas Island in 1958. The judges argued that after Hiroshima and Nagasaki in August 1945, no one could have any doubts about the effects of
nuclear weapons on human health. They said, in part: "From the outset it
was known that not only were nuclear weapons capable of causing the immediate deaths of large numbers of people but also that they could, in the
long term, have serious effects on the physical integrity and health of those
exposed to them, whether directly or indirectly, from near or from afar."[90]
To support their contention, the judges drew on two reports that the
British government had produced during the case. The first report said
that the British government was interested "in the effects of nuclear explosions on personnel and equipment" and that the government was interested in these effects "with and without various types of protection."[91]

The third source of information is the evidence presented by Daryl
Richard (Ric) Johnstone in the case *Johnstone v. Commonwealth 1988* and
the legal judgments that flowed from that case. His case is a remarkable
one—he took more than twenty years to exhaust his legal options and to

win compensation. Ric Johnstone worked at Maralinga in 1956, and as a consequence of ill health he attributed to the effects of atomic blasts, he sued the Australian government for negligence. In 1988 the NSW Supreme Court found that the Commonwealth of Australia had "a direct duty of care" to the plaintiff Johnstone and awarded him $867,100. Johnstone is the only person to succeed in litigation against the Commonwealth government in a case of this kind.[92]

Johnstone was discharged medically unfit from the army in 1958. He experienced nausea, vomiting, and diarrhea after the Maralinga tests and was treated for radiation poisoning.[93] While his experience may not replicate that of every participant at Maralinga, his testimony contradicts the assurances given on safety by Penney, Beale, and Titterton—Penney in 1956 had referred to the "outstanding importance" of safety measures and the "meticulous care" that would accompany the safety of the tests;[94] Beale had used the phrase "complete safety" on four occasions in an article on the role Australia would play in the British tests;[95] and Titterton, a member of the Australian Safety Committee, which oversaw the blasts, asserted that it was "absolutely certain" that no deaths or injuries would result from blasts, radiation, or fall-out."[96]

There are three important elements of Johnstone's testimony. First, despite safety precautions—decontamination processes such as vacuuming clothing, scrubbing his body with coarse brushes, and being tested by machines for radiation—Johnstone became ill. Second, Johnstone was given no briefings on hazards of watching atomic explosions or working at ground zero, and while he was provided with a film badge to indicate changes in radiation levels, his badge was never checked nor were measurements taken. He was told that the dosimeters had faulty batteries and the readings they obtained were thus invalid.[97]

Third, Johnstone had doubted in 1956 that he was being used as part of an experiment. He worked as a motor mechanic and part of his duties included the servicing and maintenance of vehicles, in particular those used as part of the Target Response Trials. He only concluded that he was part of an experiment or test when his superiors kept changing the clothing that he and other servicemen were to wear while working on vehicles in the Target Response areas. These vehicles never worked and required him to remove his gloves to undo small items such as wing nuts, and to take off his head gear because of the stifling desert heat.

The fourth source of information is the various reports that the British government presented in evidence. The reports from the AWRE, all relating to Operation Buffalo, are especially revealing since they contain direct and indirect references to "experiments," the use of the IF to obtain information, and uncertainty about the accuracy or efficacy of measurements. One report, *The Effect of the Thermal Radiation from a Nuclear Explosion on Service Uniforms*, says in part in its introduction, "The object of the experiments was to compare the effects on clothed men with the effects predicted from observations made on small fabric specimens."[98]

Another report on decontamination said in part, "Owing to the large numbers [of IF members] involved and the necessity for gaining experience of decontamination in the field . . . " and referred to a questionnaire IF members were required to fill in after proceeding through the decontamination centre.[99] Other reports talked about various uncertainties that attended the trials: removal of major elements such as strontium, barium, and iodine from drinking water was described as being "not easily predictable as the actual elements, as well as their proportions will vary depending on the circumstances of the explosion (i.e., underground, ground, or air burst, terrain over which exploded etc.)";[100] the "Radiac Survey Meter No.2 was found to be a reliable instrument in assessing gamma radiation but *indicated* rather than *measured accurately* the beta radiation hazard";[101] and the selection of a scientific program of experiments which was "guided by the principle that attention should be confined to questions of importance from the view-point of civil defence."[102]

Moreover, reports by members of the IF seriously challenge the assertions of Penney, Beale, and Titterton. Colonel Peters called for a "more lavish distribution of dosimeters and survey meters," indicating that there may have been an insufficient number of meters. He also recommended all future IF members should be trained in their use, indicating that not all IF members had been trained to read their meters accurately.[103] Colonel WAR Ross called for greater emphasis to be paid to "Personal Decontamination"[104] casting some doubt on the success of the field decontamination unit.

Brigadier L. H. Gordon in his recommendations for the training of future IF members also indicated that training and preparation for the IF tasks may not have been as successful as IF commanders indicated: "There will have to be a monitoring and decontamination drill and dirty and clean reception parks. There will have to be training in handling contaminated material. The psychological fear of radioactivity must be overcome and confidence established that it can be handled in safety provided the right methods are used. Drills will have to aim at handling large quantities at speed and work study will be necessary to establish efficient methods."[105]

The fifth source of information that challenges the many assertions of the bombers is the evidence presented at the Royal Commission. It concluded that the consistent reassurances of safety and risk were in fact deceitful and that the activities at the test sites were far more complex than authorities had previously reported. It also concluded that while radiation protection standards "were reasonable and compatible with the international recommendations applicable at the time," the bombers had deceived the Australian government and, indeed, many of the participants, as well as exposing them to poor monitoring and unnecessary risks.

The Indoctrinee Force Assessed

The real purposes of the Indoctrinee Force may never be known. The collection of information about the effects of radiation was inconsistent, the testing procedures to ascertain attitudinal changes were flawed, and experiments to test the effectiveness of different types of clothing also seem inconclusive. Authorities did not provide an extensive definition of "indoctrination" and, while in hindsight the inculcation of senior officers into the experience of an atomic blast by watching and then recounting their experiences may seem naive and simplistic, this assessment need not eliminate other explanations.

As Johnstone asked, "Surely there was more to it than all this?"[106] Indeed there was. Although this was not a clinical experiment and it was not the use of body as we have seen in other chapters, the "bodies" of the members of the Indoctrinee Force *were* used in a complex experiment. The answers to the real purposes of the Force are not found in the specifics of their work in the Australian desert; they are found in the broad political and social context in which the tests took place.

Britain was committed to confirming its national strengths, and Australia had an economic vision that included atomic energy and uranium sales. Both countries had powerful and influential salesmen in Penney and Titterton, who were keen to continue their nuclear experiments. The Hiroshima and Nagasaki experiences told the world in graphic detail of the horror of atomic war. If one group of senior officers could be convinced that they had experienced an atomic blast, had visited ground zero and could report to others about this experience in positive terms, ways of addressing psychological and morale aspects could begin. Simply put, the Indoctrinee Force was a small beginning to a larger program—once the military convinced its own staff, then governments could perhaps begin the next stage, that of convincing society at large.

However, legal redress by servicemen who were at Maralinga for damage to their health continues to reveal the extent of risks entailed in this scientific and human experiment and raises questions about what really occurred at Maralinga. As one obituary of Lord Penney recorded: "So, almost at the post [the Royal Commission], a blemish appeared. . . . His extraordinary achievements meant that his country could stand tall among nations. But did he risk damaging the lives of a number of people to achieve this end?"[107]

NOTES

1. CNN, "Australian troops were exposed to radiation," Web posted at 3:19 A.M. EDT, at www.cnn.com/2001/WORLD/asiapcf/auspac/05/12/australia.nuclear /index.html.

2. A detailed listing of all newspaper and television accounts is beyond the scope of this chapter. The following references are representative of the media coverage. "UK admits troops used in N-tests," *Sunday Telegraph*, 13 May 2001; "Test cases," *The Age*, 19 May 2001; "Revealed: Proof that Britain planned to put troops at heart of A-bomb fall-out', *Sunday Herald*, 20 May 2001; and "Secret documents detail plan to use servicemen in atomic tests," *7:30 Report*, ABC television, 21 May 2001.

3. CNN (n. 1 above). These were not the first experiments carried out on military personnel in Australia. In World War II, Britain carried out chemical warfare experiments, namely mustard gas tests, on Australian servicemen. See Bridget Goodwin, *Keen as Mustard: Britain's Horrific Chemical Warfare Experiments in Australia* (St. Lucia, 1998).

4. CNN (n. 1 above).

5. With a legal authority derived from the parliament of the Commonwealth of Australia and legislation of that Commonwealth, namely, the *Royal Commissions Act 1902*, a Royal Commission can subpoena documents and witnesses, interrogate this evidence in an exhaustive manner, and report its findings to the Federal parliament. It is only responsible to that parliament. It should be noted that the power of subpoena did not extend to the United Kingdom in this inquiry. The Royal Commission produced a report of three volumes, *The Report of the Royal Commission into British Nuclear Tests in Australia* (Canberra, 1985), vols. 1–3. The findings of the Royal Commission also produced several books. See Denys Blakeway and Sue Lloyd-Roberts, *Fields of Thunder: Testing Britain's Bomb* (London, 1985); Alice

Cawte, *Atomic Australia 1944–1990* (Kensington, 1992); Robert Milliken, *No Conceivable Injury: The Story of Britain and Australia's Atomic Cover-Up* (Ringwood, 1986); J. Smith, *Clouds of Deceit: The Deadly Legacy of Britain's Bomb Tests* (London, 1985) and John Symonds, *A History of British Atomic Tests in Australia* (Canberra, 1985). See also the following works published before the report of the Royal Commission: D. Robinson, *Just Testing* (London, 1985); and A. Tame and F. P. J. Robotham, *Maralinga* (Melbourne, 1982).

6. For a detailed discussion of the expansion of human experimental in medical research after World War II in America, see David J Rothman, *Strangers at the Bedside. A History of How Law and Bioethics Transformed Medical Decision Making* (New York, 1991).

7. Robert Jungk, *Brighter Than a Thousand Suns* (Harmondsworth, 1960).

8. Symonds (n. 5 above), 402. The *Report of the Royal Commission* devotes six pages to the Indoctrinee Force. See vol. 1, 335–41. Other references can be found in Blakeway and Lloyd-Roberts (n. 5 above), 4, 30, 128, 132–34, and 172; Milliken (n. 5 above), 160, 209–10, and 214–21.

9. The former AWRE is now known as the Atomic Weapons Establishment.

10. Atomic veterans in Australia, England, and New Zealand continue to pursue legal redress of their claims. At the time of writing, there is at least one case before the Australian Federal Court. The admissions by the MoD and the production of further documents following the research of Sue Rabbitt-Roff indicate that more material may be available than has been previously acknowledged. For example, in June 2001 *The Independent* revealed that a 41–page document submitted by the MoD to the Royal Commission was originally at least 75 pages long and that many names, including men from the IF who had received more than ten times the maximum permissible dose, had been struck from the list. See *The Independent*, "UK 'tampered with evidence of nuclear tests,'" 14 June 2001.

11. This phrase is quoted in Raphael Samuel, "Local History and Oral History," *History Workshop*, no. 1, Spring 1976, 204.

12. Interview with Mr. W. M. Aird, former IF member, 6 November 2000.

13. The Liddell Hart Centre for Military Archives at King's College, London, holds the papers of Edward R. Drake Seager, who was the Indoctrinee Force coordinator. However, I was unable to access this material. It is reasonable to assume that this material would throw additional light on the work of the IF.

14. The vast literature of the bombings, dating from 1946, is beyond the scope of this chapter. However, see *The Committee for the Compilation of Materials on Damage Caused by the Atomic Bombs in Hiroshima and Nagasaki, Hiroshima and Nagasaki: The Physical, Medical, and Social Effects of the Atomic Bombings* (London, 1981) for a thorough coverage of these themes.

15. See P. M. S. Blackett, *Fear, War and the Bomb: Military and Political Consequences of Atomic Energy* (New York, 1949); and M. J. Sherwin, *A World Destroyed: The Atomic Bomb and the Grand Alliance* (New York, 1975).

16. M. Susan Lindee, *Suffering Made Real: American Science and the Survivors at Hiroshima* (Chicago, 1994), 14–15.

17. For a detailed discussion of the economic power envisioned by some Australian politicians, see Cawte (n. 5 above).

18. See Richard Rhodes, *The Making of the Atomic Bomb* (New York, 1986).

19. Lorna Arnold, *Britain and the H-Bomb* (Basingstoke, 2001) 72.

20. Margaret Gowing, assisted by Lorna Arnold, *Independence and Deterrence: Britain and Atomic Energy, 1945–1952*, Vol. 1, *Policy Making*, and Vol. 2, *Policy Execution* (London, 1974).

21. For an examination of the documents that described the relationship between Britain and the United States, see D. S. Patterson, ed., *Foreign Relations of the United States, 1955–1957: Volume XX Regulation of Armaments; Atomic Energy* (Washington, D.C., 1990); B. Cathcart, *Test of Greatness. Britain's Struggle for the Atom Bomb* (London, 1994) 20.

22. Gowing (n. 22 above), 160.

23. Ibid., Appendix 8, 194.

24. See Blakeway and Lloyd-Roberts (n. 5 above), chap. 2, pp. 10–28.

25. Arnold (n. 19 above), chap. 14, pp. 195–220. See also J. Bayliss, *Anglo-American Defence Relations 1939–1984*, 2nd ed. (London, 1984).

26. Lorna Arnold, *A Very Special Relationship: British Atomic Weapon Trials in Australia* (London, 1987).

27. Milliken (n. 5 above), 238–80, vol. 2 of the *Report of the Royal Commission* (n. 5 above), 395–415; and Symonds (n. 5 above), chaps. 20–23, pp. 478–532.

28. *Commonwealth Parliamentary Debates*, 27 September 1956, 906.

29. For an excellent discussion of the secrecy of these tests, see Blakeway and Lloyd-Roberts (n. 5 above), 59–75.

30. *Sydney Morning Herald*, 15 August 1956.

31. *Sydney Morning Herald*, 13 August 1956.

32. Blakeway and Lloyd-Roberts (n. 5 above), 104–5.

33. Ibid.; R. L. Miller, *Under the Cloud: The Decades of Nuclear Testing* (New York, 1986).

34. *Commonwealth Parliamentary Debates*, 5 September 1956.

35. Interview with Sir Mark Oliphant, former research director in physical sciences at the Australian National University, 19 August 1998. Oliphant was a colleague of Titterton, but unlike Titterton, Oliphant campaigned for peaceful applications of atomic energy and "against the policy of secrecy surrounding development of the atomic bomb" and Britain's tests in Australia (*The Penguin Australian Encyclopaedia* [Ringwood, 1990], 369).

36. Milliken (n. 5 above), 52.

37. Interview with Mr. H. Hooton, 23 May 2000. Hooton, member of the IF, stressed the secrecy of his work: "Not even the family knew."

38. For an interesting review of the work at Aldermaston, see G. Spinardi, "Aldermaston and British Nuclear Weapons Development: Testing the 'Zuckerman Thesis,'" *Social Studies of Science* 27 (1997): 547–82.

39. AWRE, *Operation Buffalo, Summary Plan, Section B29—Indoctrinee Force,* June 1956. 5. There is no explanation as to how the AWRE constructed its definition of "safe."

40. Report of the Royal Commission (n. 5 above), 1:336.

41. AWRE, *Operation Buffalo, Summary Plan, Section C1–Administration.* 61.

42. Interview with Mr. W. M. Aird, 6 November 2000. Aird was a member of

the IF who believed no member was given information on these matters: "I received no follow up material whatsoever on the work of the Force."

43. In a case before the Australian Federal Court at the time of writing, counsel for a former Royal Australian Air Force officer has accused the Australian government of "devious" and "false" information about the safety indications that photo badges allegedly provided (*Sydney Morning Herald*, 5 June 2001).

44. E. R. Drake Seager, *Operation Buffalo: The Indoctrination of Serving Officers*, AWRE Report No T1/93.

45. Ibid.

46. See various files in the papers of Sir Ernest Titterton, Australian Academy of Science Archives, Canberra. Titterton leaves the reader in no doubt that Maralinga was a valuable site for wide-ranging experiments.

47. Report of the Royal Commission (n. 5 above), 348.

48. Ibid., 325–26.

49. Peter McNeill, *The Ethics and Politics of Human Experimentation* (Melbourne, 1993), 30.

50. House Subcommittee on Oversight and Investigations, *"The Forgotten Guinea Pigs": A Report on Health Effects of Low-Level Radiation Sustained as a Result of the Nuclear Weapons Testing Program Conducted by the United States Government* (Washington, D.C., 1980).

51. Robert Walgate, "Moving Standards in 1950s," *Nature* 213 (17 January 1985): 175. Rotblat has written extensively on nuclear disarmament. See Joseph Rotblat, ed., *Scientists, the Arms Race, and Disarmament: A UNESCO/Pugwash Symposium* (London, 1982); and Joseph Rotblat, *Scientists in the Quest for Peace: A History of the Pugwash Conferences* (Cambridge, MA, 1972). He was awarded the Nobel Peace Prize in 1995.

52. The ICRP was established in 1928 by the International Society of Radiology to provide recommendations and information on radiation protection. It was restructured in 1950 to encompass radiation use outside medical applications.

53. The Titterton papers, for example, stress the "safety" of the tests.

54. Interviews with Mr. H. Hooton and Mr. W. M. Aird, both of whom confirm this assessment.

55. See J. Weisgall, *Operation Crossroads: The Atomic Tests at Bikini Atoll* (Annapolis, MD, 1994); and Barton Hacker, *The Dragon's Tail: Radiation Safety in the Manhattan Project, 1942–1946* (Berkeley, CA, 1987).

56. Department of the Scientific Adviser to the Army Council, Army Operational Research Group, Report No. 9/57, *The Value of Live Indoctrination at a Nuclear Weapon Trial (Operation Buffalo)*, November 1957. (Hereafter called the "Holman report.")

57. See AWRE, *Operation Buffalo, Summary Plan, Section B29* (n. 39 above); document entitled *Copy Indoctrinee Force Commanders Report*, no date; and E. R. Drake Seager's testimony to the Royal Commission, at RC 325 to show the absence of definitions.

58. Holman report (n. 56 above), Abstract 1.

59. Ibid. The report does not describe how the control was selected, how many men made up this group, and what materials were withheld from their work.

60. Ibid.

61. This report is Appendix G of Holman report (n. 56 above).

62. Lt.-Col. W. N. Saxby, *Indoctrinee Force Commander's Report*, Buffalo Trials (hereafter IFCR), December 1956, 1.

63. IFCR, Annexure G.2.

64. IFCR, Annexure L 1.

65. IFCR, Annexure H. 2.

66. IFCR, Annexure I. 2.

67. See Allan Mazur, "Disputes Between Experts," *Minerva* 11 (1973): 243–62; Dorothy Nelkin, "Scientists in an Environmental Controversy," *Science Studies* 1 (1971): 245–61; and Dorothy Nelkin, "The Political Impact of Technical Expertise," *Social Studies of Science* 5 (1975): 35–54.

68. Commonwealth Archives, Canberra. Transcript of the Royal Commission, T4335.

69. D. S. Patterson, ed. (n. 21 above), 75.

70. *Sydney Morning Herald*, 31 July 1956.

71. *Sydney Morning Herald*, 28 September 1956.

72. Letter in Titterton Papers, Australian Academy of Science.

73. Titterton Papers, Series 13–Atomic Weapons Testing Safety Committee, Correspondence-Section 3, Titterton to Beale, 14 October 1957.

74. Titterton Papers, Series 22–Royal Commission into British Nuclear Tests in Australia (McClelland Royal Commission), Section 39, "Typing for EWT"— miscellaneous material, Titterton to Atomic Energy Research Establishment, 18 October 1956.

75. Ernest Titterton, *Facing the Atomic Future* (Melbourne, 1956), 284 and 183.

76. Titterton Papers, Series 22 (n. 74 above). Author's emphasis.

77. Titterton Papers, Series 22–Royal Commission into British Nuclear Tests in Australia (McClelland Royal Commission), Section 39, "Typing for EWT"— miscellaneous material, Titterton to Atomic Weapons Research Establishment, 9 August 1960.

78. Titterton Papers, Series 31–Australian Broadcasting Commission, Section 1 Correspondence and Scripts of Talks-Printed and Handwritten Sheets, 1956. Titterton to A. Carmichael, Director of Talks, 8 August 1956.

79. R. Bertell, *No Immediate Danger: Prognosis for a Radioactive Earth* (London, 1985), 275. Interview Friday 17 July 1998 with Associate Professor Brian Martin, University of Wollongong, a former researcher at the Australian National University who clashed with Titterton on several occasions.

80. Australian Archives, Canberra (Series A1209/23), Item number 57/4467–Atomic Tests Maralinga-Biological Aspects & Use on Animals. Press statement by H. Beale, 24 August 1956.

81. Ibid.

82. Ibid.

83. *Arizona Republic*, 18 August 1995.

84. H. Wasserman and N. Solomon, *Killing Our Own: The Disaster of America's Experience with Atomic Radiation* (New York, 1982) and the Advisory Committee on Human Radiation Experiments, Final Report (hereafter ACHRE Report), chap.

10, "Human Research at the Bomb Tests" (Oxford, 1996), 286–90.

85. ACHRE Report, 286

86. The Armed Forces Medical Policy Council was the Secretary of Defense's principal medical advisory body.

87. Richard L. Meiling to the Deputy Secretary of Defense et al., 27 June 1951 ("Military Medical Problems Associated with Military Participation in Atomic Energy Commission Tests") (ACHRE No. DOD-122794–B), 1. Quoted in ACHRE Report, 286.

88. Ibid.

89. This reference draws on a transcript of meeting quoted by ACHRE to discuss the "Psychological Problem of Crew Selection Relative to the Special Hazards of Irradiation Exposure." See chapter 10, n. 3.

90. European Court of Human Rights, Case of McGinley and Egan v. The United Kingdom, (10/1997/794/995–996), 9 June 1998, 33.

91. Ibid., 33, notes 1 and 2.

92. See Supreme Court judgments, No. 14919 of 1979 (11 November 1988, 21 November 1988, 5 December 1988 and 6 April 1989.)

93. Interview with Ric Johnstone, 17 July 2000.

94. See report of Penney's press conference, *Sydney Morning Herald*, 15 August 1956.

95. See his article, "Why Australia Provides the Site for Atomic Tests," *Sydney Morning Herald*, 13 August 1956.

96. See article, "Atom Tests Safe says Prof. Titterton," *Sydney Morning Herald*, 31 July 1956.

97. Ibid.

98. AWRE, Report No. T11/58, 3.

99. AWRE, *The Construction and Operation of a Field Radiological Decontamination Centre*, Report No. T1/57, 3.

100. AWRE, *The Decontamination of Radioactively Contaminated Drinking Water in the Field*, Report No. T4/57, 3–4.

101. AWRE, *The Measurement of Radiation Dose-Rates from Fallout*, Report No. T40/57. Author's emphasis.

102. AWRE, *Interim Report, Target Response-Biology Group*, Report No. T18/57, 11.

103. IFCR (n. 62 above), Annexure G, 3.

104. Ibid., Annexure D, 2.

105. Ibid., Annexure H, 2.

106. Interview with Ric Johnstone, 17 July 2000.

107. D. Andrews, ed., *The Annual Obituary* (London, 1992), 166.

Part III: Whose Body?

CHAPTER SEVEN

Injecting Comatose Patients with Uranium

America's Overlapping Wars against Communism and Cancer in the 1950s

Gilbert Whittemore and Miriam Boleyn-Fitzgerald

Medical experimentation often puts physicians into the dual roles of care-giver and scientist. At the beginning of the twenty-first century, we see ethical concerns over a third role—entrepreneur—which seeks profit, not health care, as its prime goal. But profit-making enterprises are not the only institutions to impose a nontherapeutic purpose onto a medical experiment; governments also do so. The World War II injections of uranium and plutonium into unknowing human subjects as part of the U.S. atomic bomb project are now well known. In 1994 investigative reporter Eileen Welsome provoked a public furor by identifying some of the hitherto anonymous subjects and telling their personal stories. One result was the appointment by President Bill Clinton of the Advisory Committee on Human Radiation Experiments (ACHRE). Working from 1994 to 1995, ACHRE published a voluminous report and appendixes. Less visible but perhaps more important to historians, ACHRE also obtained from a wide variety of government agencies the declassification of many documents now contained in the ACHRE archives.[1]

"Dual purpose" experiments, those serving goals of national defense as well as peacetime health care, formed a large part of the human radiation experiments investigated by ACHRE. Studying these reveals how complex an experiment can become when it serves two masters. In addition to the obvious potential for ethical conflicts, scientific and institutional choices become more convoluted. This chapter examines one such experiment from the height of the cold war. The examination reveals the dual culture underlying such experiments and illustrates the need to access government archival material to understand such research.

From 1953 to 1957, in an experiment jointly conducted by the Atomic Energy Commission (AEC) and Dr. William Sweet at Massachusetts General Hospital (MGH), eleven terminally ill brain cancer patients were injected with uranium.[2] Termed the "Boston Project," the experiment measured where within their bodies the uranium settled, both to explore a potential therapy for brain cancer and to model the metabolism of uranium for industrial health standards. Newell Stannard, in his massive history of radiation biology, remarked: "Although there is no evidence that these administrations were in any way harmful or ill-considered at the time, it is difficult to imagine a present day 'human use' committee approving them."[3] In its 1995 final report, ACHRE stated: "Unless these patients, or the families of comatose or incompetent patients, understood that the injections were not for their benefit and still agreed to the injections, this experiment also was unethical. There was no justification for using dying patients as mere means to the ends of the investigators and the AEC."[4]

In what way were the subjects of this experiment "used bodies"? First, their bodies were used to obtain information useful for industrial health practices in the production of nuclear weapons. Second, they were terminal cancer patients whose bodies were used in a preliminary test of a potential therapy but without any hope that this test would be of therapeutic value to them. Third, most were comatose, posing the difficult question: When does a person become simply a "body," no longer having even the hope of recovering intellect or will, kept alive only with the technology of an advanced hospital? Fourth, the use of their bodies bridged two worlds—the secret world of nuclear weapons production, and the elite world of medical research—allowing a direct contrast of two professional cultures.

The War on Cancer

This one experiment was a small battle in a much larger war. Brain cancer, specifically glioblastoma multiforme, had long been regarded as terminal. The onset of this cancer was dramatic and tragic. Sweet told of one young, active woman who got into her car one day and closed the door to drive off. The door would not shut. Looking down, she realized that her right leg was still hanging outside the car threshold. "This is not right," she reported to her doctor. "I should have known where my leg was." Within a few days a brain tumor had been diagnosed, and despite a combination of surgical removal—the conventional treatment at the time— and Brookhaven's experimental boron-neutron therapy, she soon died.[5] Doctors today describe the disease as one that still is tragic, terrible, and untreatable.[6]

Soon after the end of World War II, Sweet, a young Boston neurosurgeon, saw ahead of him a career developing surgery to remove cancers. But the secrets of the atom, credited with winning the war against Japan, were now being promoted as the wonder weapon that might also win the war against cancer. Sweet frequently provided a ride for Professor Baird Hastings, chair of the biochemistry department at Harvard Medical School, who did not drive. In the car Hastings suggested that Sweet might examine the use of the newly available radioisotopes for cancer therapy as an alternative to the scalpel.[7] Sweet was intrigued enough to return a grant from the Rockefeller Foundation to develop seterotactic neurosurgery and instead to seek resources from the AEC to investigate radiotherapy.[8]

One type of radiotherapy was the introduction of radioisotopes into the body. Dramatic success had been achieved during the war with the use of radioiodine for thyroid therapy, taking advantage of the body's innate tendency to concentrate iodine in the thyroid gland. Could one find other radioisotopes that were naturally concentrated in a specific organ? If so, could these be used to attack a tumor in that organ?

Sweet's specific interest was brain cancer. His initial use of a radioisotope was to locate tumors during surgery. Phosphorous-32, the very first radioisotope that Sweet tested, had "an extraordinary tendency to concentrate in brain tumors as compared with normal brain."[9] Sweet used phosphorous-32 to develop a technique using radiation detectors on probes to locate the boundaries of a tumor, allowing more precise surgery.[10] He sought, but would never find, another radioisotope that not only concentrated as well in brain tumors but also could deliver, from inside the tumor itself, a therapeutic dose of radiation. "We were looking for the magic compound that would do it the way phosphate did, which we never found."[11]

A second type of radiotherapy was applying an external beam of radiation to the cancerous tissue. Before the war, x-ray machines and vials of radon gas had been the primary sources. By the end of the 1930s, cyclotrons were also being built to produce beams for therapy. But external beams could not be as finely aimed as radioiodine. Essentially, one was aiming a spray of bullets at the body and hoping that, by focusing the spray and choosing bullets with precisely the right penetrating power, only the target tumor would be damaged.

After the war, nuclear reactors presented another source of radiation—this time in the form of neutrons. The primary advantage of neutrons was that they could transform a non-radioactive isotope into a radioactive one. Within a year of his initial use of phosphorous-32 to locate tumors, Sweet read an article on the destruction of lily bulbs by interaction between slow neutrons and boron.[12] Sweet and others speculated that one might be able to combine the two types of therapy. Could they find a non-radioactive isotope that would be concentrated in a brain tumor and then make it radioactive with an accurately focused beam of neutrons?

The key would be an isotope that passed through the blood-brain into the tumor faster than it spread through the remainder of the body.[13] One promising isotope was boron-10. When struck by a beam of neutrons, boron-10 would absorb a neutron to become radioactive boron-11. Sweet believed that boron traveled into the brain faster than throughout the remainder of the body; if so, the neutron beam could be applied during the period in which the brain had this higher concentration of boron. Some were skeptical about the entire approach. The eminent Shields Warren, M.D., former director of the AEC's Division of Biology and Medicine, shocked Sweet by bluntly commenting, "I think it's a lousy idea. It will never work."[14]

Despite Warren's skepticism, the first experiments were conducted at the AEC lab at Brookhaven, New York. Almost half a century later, Sweet vividly recalled: "They had a coffin-like excavation at the top of the shield of the reactor. That was all cranes and all this heavy machinery. The patient lay down in this area, about half the size of this table. We then all retreated way back to the control room."[15] Sweet scoffed at the concern of the security officers that the patients did not see secret details of the reactor. The patients could not see the "loading face" of the reactor, and even if they had, they "wouldn't know the loading face from Aunt Mamie. These were sick people."[16] As late as 1995, hanging on the walls of the lobby at

Brookhaven was a painting of an idealized setting of the patient, with head exposed, being placed over an opening at the top of the reactor, with uniformed medical staff at controls behind a shielding wall, a dramatic image of science applying the power of the atom to the cure of an individual patient.

The initial experiment was conducted on ten patients.[17] Two major difficulties arose: detecting precisely when the boron was highly concentrated in the tumor, and focusing the neutron beam accurately. The experiments ended when, in Sweet's words, "to our horror, on patient number 21, she got a very much larger dose of radiation, and died in three weeks of a radiation injury."[18] After this failed dream, Sweet and others began looking for other radioisotopes to attack brain tumors.

The War on Communism

The AEC was arming for a different war, producing atomic weapons using "special nuclear materials" that were highly guarded. Industrial health was a continuing problem, especially the inhalation or ingestion of radioactive material. Setting a standard for air and water concentration required determining where within the body each element settled. An element might do greater harm if the body concentrated it in one crucial organ instead of distributing it evenly throughout the body or, better yet, rapidly eliminating it.

At issue was the perennial tension between safety and productivity. The health physics department would remove a worker from uranium operations when his cumulative internal exposure, as calculated from uranium measured in urine, exceeded allowable standards.[19] By 1956, about thirty-five workers had been so removed at the Y-12 plant in Oak Ridge, Tennessee.[20] Out of a work force of about one hundred, this was a sizeable loss of manpower. Many years later the health physicist S. R. Bernard recalled: "This had management all upset. Jesus, taking production workers away from their work, stuff like that."[21] The health physicists faced the problem of convincing a skeptical audience that long-term harm was occurring, even though no harm could yet be observed. "They [management] listened to us. But they would argue with us and say, 'Well, for godsakes, Joe over there, he's been working at this for five years and he's got ten kids,' or something like that. We would tell them that it's the long-term effect that we're worried about. They would listen and argue back, pooh-pooh it, that sort of thing."[22]

The problem was lack of human data. Animal data were easily produced, but their application to humans was ambiguous, given the observed variation among species. "Mice tolerated 100 times more [radiation exposure] than rabbits, and certain strains of mice, 200 times more."[23] Which species, if any, was an adequate model for humans?

The lack of human data was due to an ethical problem. Studies on the metabolism and toxicity of radioisotopes would not benefit the human subjects of such experiments. During the war, uranium, plutonium, and polonium had been injected into humans subjects, at times without their knowledge. These experiments prompted the 1994 investigation by ACHRE.[24] One conclusion of ACHRE was that these experiments had been halted in 1946 precisely because of concern over the ethics of such work. Even if he had wished to, it is unlikely that Karl Morgan, head of health physics at Oak Ridge, could have undertaken similar experiments solely to obtain the data he needed. Instead, he established a long-term research effort to collect data on the metabolism of each of the elements. Trained as an astrophysicist, he was amazed to find, on his venture into the medical literature, how little was actually known about the distribution within the body of the various elements. The collection became known as the "MPC [maximum permissible concentration] library." He intended to establish exposure limits by developing mathematical techniques that would combine knowledge of the radioactivity of each isotope with knowledge of how the body metabolized it.

Possible sources of data for many isotopes were medical research experiments. Complementing its weapons development, the AEC vigorously promoted the use of radioisotopes in medical research, under careful control. But the isotopes of greatest concern at Oak Ridge were those used in atomic bombs: uranium, plutonium, and polonium. Such "special nuclear materials" were closely guarded and not part of the AEC's civilian medical radioisotope program.

A particular problem arose for the health physicists when, after convincing management to take steps to deal with the chemical toxicity of natural uranium, they had to return to management with renewed concern over the radiation effects of uranium 234—about 1 percent of "product [i.e., weapons] level" enriched uranium, but active enough to be a special problem. In an attempt to gather data to support their position, the Oak Ridge health physicists consulted with Harold Hodge at the University of Rochester. Hodge was active in many areas, including long-term animal experiments on radiation effects and chemical toxicity.[25]

Combining Fronts in the Two Wars

Could the war on brain cancer and the war on communism be fought on a single front? Hodge put Oak Ridge in contact with Sweet. S. R. Bernard later recalled: "We traveled to Rochester and would get help. And then Harold Hodge said, 'I tell you what. There's a guy by the name of Bill Sweet, who wants to use uranium, enriched uranium, in brain tumor patients.' He says, 'Why don't you fellows contact him and see if you can work up something with him and get some excretion data to check your procedure out with.'" [26] By then, Sweet had given up on the boron-neutron capture therapy done at Brookhaven and "wanted to try uranium," Bernard recalled, "to see if he could get a small fission reaction in brain tumor cells. But he had to know if enriched uranium concentrated in the tumor cells and if it did how fast relative to normal cells and things like that. He was agreeable to injecting enriched uranium or [uranium] 233 into the patients, into the tumor patients anyhow."[27] In 1953, on a field trip to MGH, the AEC's Advisory Committee for Biology and Medicine visited Sweet.[28] Soon after the Advisory Committee's visit, Sweet and his research assistant, Luessenhop, met with Struxness and Bernard of Oak Ridge.[29]

Sweet's program dovetailed with AEC needs. Sweet required measurements of where in the body uranium was distributed, exactly the type of data the AEC needed to develop industrial health standards. How much uranium was retained in the body could be measured by comparing the amount injected with the amount excreted. And excretions could be more reliably measured with comatose patients.[30]

The use of terminal patients also would allow additional data to be gathered from autopsies. Oak Ridge was especially interested in obtaining tissue samples from autopsies so that the distribution of uranium throughout the body could be measured.[31] Samples would include "a whole femur, a whole kidney, and large pieces of liver and spleen, and a sagittal half of kidney, in addition to pieces of all other organs."[32] As one AEC health physicist wrote, the experiment "provides a wonderful opportunity to secure 'human data' for the analysis and interpretation of industrial uranium exposures."[33]

The experiment seemed rushed, even opportunistic. No animal experiments preceded the human experiments. The published report refers only to "careful theoretical evaluation of the clinical use of fissionable uranium."[34] The theoretical evaluation apparently consisted of noting that uranium-235, once activated by neutron capture, would release more energy when decaying than boron or lithium, and that "the tremendous size of the uranium atom suggests a more complete exclusion by the blood-brain barrier and hence a more efficacious differential concentration between tumor and normal brain."[35] No data or citation was given for this suggestion concerning the permeability of the blood-brain barrier and the size of single atom, a different matter than the permeability of the barrier and the size of large, complex molecules. In contrast, plutonium studies began with mouse experiments at Brookhaven, and never proceeded to human experiments.[36] Planning proceeded at meetings between health physicists from Oak Ridge and Sweet's staff.[37] No financial arrangements were made; Oak Ridge and MGH would each cover their own costs.[38] This reduced red tape, which also reduced oversight and formal accountability.

Combining armies is never simple. The two goals of the experiment are clear from the plans for data collection. Sweet's MGH team provided Oak Ridge with data needed for determining how quickly the body excreted a known amount of uranium, a crucial step in establishing industrial health standards.[39] By September 1953, Luessenhop agreed that MGH "will be able to completely comply" with AEC requests for additional tissue samples.[40] So far as Sweet's brain tumor research was concerned, only analyses of cerebralspinal fluid and excretions was needed.[41]

But although the two data collection efforts were compatible, there was an incompatibility concerning the level of dose. The AEC desired data on small doses, approximating occupational exposure. Sweet believed that the highest dose was desirable to maximize the difference between tumor and normal brain. For him, therefore, the initial step was "to investigate the toxicity of uranium in man with a view to determining the highest intravenous injection dose."[42] As the experiment progressed, the dose given patients was increased to meet Sweet's goal, even though, for the AEC's goal, it was more important to study smaller doses.[43]

Long-term goals also diverged. The AEC was concerned with the occupational hazards of many radioactive substances, not just uranium. As the experiment progressed, health physicists at Oak Ridge recommended that "the possibility of extending this study to plutonium, thorium, and the critical fission products" be considered.[44] A recommendation was also made for additional human studies on inhalation as well as injection of uranium.[45] No argument was made that studying these additional elements or inhalation would help Sweet find a treatment for brain tumors; the sole criteria were the industrial hazards in the AEC's plants.

Using Uranium

Ironically, the use of tightly controlled uranium allowed the experimenters to sidestep the normal oversights of the AEC's radioisotope program. The AEC's primary concern was with the care of the uranium, not the human subjects. Soon after Luessenhop's letter to Oak Ridge, Struxness formally requested "SF material," explaining that "Dr. Sweet's program is actually directed toward the therapeutic properties of fissionable materials in the treatment of brain tumors."[46] The use of the word "actually" is intriguing, suggesting that Sweet's goal, by itself, would not justify this request. This could well have been the case, since purely therapeutic uses of radioisotopes were overseen by the AEC's Division of Biology and Medicine and were subject to approval by its Subcommittee on Human Applications. But the uranium for the Boston Project was not treated as a delivery outside the AEC subject to such oversight; it was regarded as only an allocation within the AEC of special nuclear material.

Struxness went on to describe the second purpose of the experiment, which brought it within the purview of the health physics department of the Y-12 weapons plant: "However, by judicious design of the early experiments it will be possible to use the results of this study in the evaluation of internal exposures to uranium. Such an evaluation will find direct application by this department in its assessment of the internal exposures of plant employees."[47] Within a week, Carbide and Carbon Chemicals Company, the contractor that would furnish the uranium-233, sought formal approval from the AEC office at Oak Ridge. But Carbide made no mention of Sweet's goal; only the determination of "the distribution of uranium in the tissues and organs of humans exposed to this material" was listed as the purpose of the experiment.[48]

Preparation of the uranium began on October 30, 1953.[49] Oak Ridge labs worked late into the night. The first solution was finally completed at 9:00 P.M., at which point "E. G. Struxness and W. C. Emerson, under the escort of two Y-12 guards, journeyed to X-10 to have the solutions prepared by the Packaging Dept. for shipment."[50] Meanwhile, the filter paper used in preparing the solution was recovered, through the use of "high range" detectors, and approximately one-half Mg of uranium-233 extracted from it. This careful documentation of the armed escort (while still within the guarded grounds of Oak Ridge itself) and the recovery of tiny amounts of uranium-233 from filter paper contrasts with the scant documentation later at MGH of subject selection and consent.

By 4:30 A.M. the shipment was prepared. Transportation from Oak Ridge to Boston had to be approved by the Oak Ridge Security Department.[51] Formal responsibility for the material was transferred from Oak Ridge to the AEC regional office in New York, with a "Research Issuance" for Sweet.[52] The AEC expected that its own rigor would be continued at Massachusetts General. But shipped by air to Boston, the uranium escaped the government world of armed guards and arrived in the civilian world, where the package was picked up at the airport by a lone lab assistant.[53]

It soon became clear that the culture of MGH did not conform to the AEC's stringent security and record-keeping requirements. Struxness at Oak Ridge recommended that he assume responsibility for the uranium-233 "within the limits imposed by standard accountability procedures." As he put it: "Dr. William H. Sweet and his associates, never having handled fissionable materials before, are not familiar with the requirements and are ill equipped to comply with these requirements."[54]

Using Bodies

Just as producing uranium was a routine task for the weapons labs at Oak Ridge, obtaining human subjects was a routine task for a research hospital. In August 1953 Luessenhop reported that "considering the patients we have available, it appears that we will be able to extend the experiment over a period of many months."[55] But like the requisition of uranium-233, this too required some formalities. On November 3, 1953, just four days after the uranium had arrived at MGH, the experiment was approved by Isotope Committee of MGH.[56]

The goals of the experiment were "to procure data on metabolism and toxicity for AEC and differential conc [*sic*] in brain tumors with view to later using U-235 and subjecting tumor to neutron flux."[57] Clearly, the subjects in this study—described as "about 5 pts. [patients]—terminal with gliomate"[58]—would not themselves benefit. Only later might uranium injections be combined with neutron beams to attack brain tumors. The dosage of uranium would vary with patient weight and be administered intravenously over two minutes. At the time, deposition of uranium in the skeleton was regarded as the greatest hazard, based on the tragic experience of the radium dial painters, where radium had become permanently lodged in bone. The application calculated that the dose to a patient would be 2.12 rem/week to bone and noted in its last sentence: "This exceeds max. permissible exposures of 0.3 rem/wk but pts. are terminal."[59]

The primary concern of the committee was not protection of the subjects but the safety of the researchers. For example, a group from Deaconess Hospital would monitor the lab weekly for radiation; the uranium itself would be kept in a box made of lead bricks.[60] But once the hospital's Isotope Committee approved the application, there was no further monitoring of actual dose rate. One patient may have received up to 50 milligrams of uranium, twenty-five times more mass than that stated in the application.[61]

Not until November 1953 did a patient become available who was considered by the MGH team to be "most suitable."[62] But terminal patients were sought by several research groups, and someone else had reached this patient first. "However, today the pt. was given 2 millicuries of P-32 by another group. . . . If you feel that the presence of this isotope may interfere with the electroplating method of uranium analysis, we will wait for another patient."[63] Oak Ridge did not wish to wait, and the experiment proceeded, although the presence of P-32 did interfere later with tissue analyses. By November 18 the first samples were on their way to Oak Ridge.[64]

At first it seemed that there would be a steady stream of suitable patients. On November 30, MGH reported that "At the moment there is a strong possibility that we will have another patient sometime next week."[65] Gradually, this early optimism faded. By April 1955, Sweet's assistant was apologizing to Oak Ridge: "We are all sorry that we have had so much trouble finding a suitable U patient."[66] There is no documentation indicating whether the "trouble" was simply a lack of patients diagnosed with brain tumors, competition with other research programs, or reluctance on the part of patients or their relatives to participate. Nor were all the subjects from Massachusetts General: one died at Boston City Hospital; two died at Holy Ghost Hospital;[67] one was discharged home and later returned to the Veterans Administration Hospital.[68]

How did the experiment appear to the subjects and their families?

Records indicate little. There is no documentation of patient permissions except for one instance of a signed permission to publish a photograph of a patient.[69] One patient, who recovered enough to become ambulatory and was discharged from the hospital for a time, cooperated by continuing to supply urine samples.[70] But at least one family refused permission for an autopsy on one of the subjects, a 63–year-old patient at Holy Ghost Hospital.[71]

In 1995 Sweet maintained that permission to conduct an experimental procedure was always obtained from family members in addition to the patient: "We relied on communication with the family, because even an individual who is much of the time alert and [has the] cognitive function to understand yes, it's a dangerous situation, we had no way of proving that. We felt that the next of kin responsible—relative or relatives—had to be the ones who were told what we were up to."[72] It is important to remember that such consent was coming after the patient, according to Sweet, had already been diagnosed as terminally ill by their own physicians. "They were all extremely grateful, and assumed that what doctors already had told them, that the situation was totally hopeless. If you're willing to spend some time trying to help them out, they're just too grateful for words."[73]

What is not known is the degree to which Sweet, by portraying his research as "helping them out," may have been promising more than these nontherapeutic experiments could ever provide. One witness before the Advisory Committee, who volunteered at MGH as a 17–year-old, recalled her conversations with the hopeful family when visiting one of the subjects: "This time, his wife and adult son were visiting, weeping softly in the darkened room. I explained myself and the peach. Mrs. Lefton [the subject's wife] was kindly. She said he was too sick to eat the fruit, but that because he was being treated by Dr. Sweet, they had hopes he would recover. Indeed, she remarked, after each 'treatment,' he did seem a little better, but it didn't last, at least so far."[74] But her conversations with staff were less optimistic: "At Jacob's third 'treatment,' I repeated Ms. Lefton's comments about his temporary improvement, and asked what the future held for Jacob. Someone explained that Jacob's increased responsiveness after each procedure was, in fact, due to the spillage of his brain and tumor tissue, which temporarily lessened pressure on his brain. Jacob was not receiving 'treatment.' There was no treatment. The experiment was not for the benefit of the patient."[75]

An even more serious ethical problem was misdiagnosis. "[A]t least one patient injected with uranium did not have a brain tumor at all. An uniden-tified male, identity and age still unknown at the time of his death, became Boston Project subject VI when 'he was brought to the Emergency Ward after being found unconscious. . . . No other information was available. According to his autopsy report, this patient was suffering from subdural hematoma—a severe hemorrhage—on the brain."[76] Such a patient could not provide any useful data for a brain tumor study. Moreover, how could consent have been obtained for an unconscious patient whose identity was unknown?

Consequences of Haste

One consequence of the haste with which the experiment was undertaken was the failure to perfect basic lab techniques in animal experiments. Difficulties arose in data collection. A crucial aspect of the experiment was measuring exactly how much normal uranium and how much uranium-233 were injected. This would allow researchers to measure radioactivity from uranium-233 in excretion and tissue samples and then calculate how much total uranium was excreted, how much remained in the body, and where within the body uranium was deposited. But for the very first subject, the two solutions—one of "normal uranium" and one of uranium-233—were mixed. It was not possible to determine from the residue how much normal uranium and how much uranium-233 remained and, from this, how much of each had been injected.[77]

Also, on the first and second patients, the stopcocks in the injection tube malfunctioned; the procedure was later changed to using a syringe to inject the solution into the rubber intravenous tubing.[78] With no accurate actual measure of the amount of uranium-233 injected, the researchers estimated by calculating backwards from the excretion rates—using the very numerical models that the experiment was supposed to be testing. On later patients a "dummy injection" into a flask was done first, and the amount in the dummy sent to Oak Ridge to provide a more accurate measure of the actual injection. Even this did not always work. On the last three patients, the amount of uranium actually excreted by the patient exceeded the uranium in the dummy injection.[79] The published study ignores this and reports that the dummy injection technique "accounted for any volume errors as a result of inaccurate markings on the syringe."[80]

The next crucial step was obtaining useful samples after the injections.

But contamination appears to have occurred with at least the first patient, since "identical" samples yielded differing results. Oak Ridge and MGH worked to improve the experimental techniques.[81] The initial ashed samples from the autopsy of the first patient apparently were contaminated with flecks from the interior of the furnace, which was replaced.[82] Hoods and ducts in the lab at MGH were "damaged," perhaps by radioactive contamination.[83]

For accuracy in data collection, the researchers clearly preferred hospitalized patients, especially comatose patients on catheters. Patient cooperation was not always adequate. Collection of urine ceased in the case of one patient when discharged from MGH to Holy Ghost Hospital.[84] Another patient recovered to the point of becoming ambulatory and was discharged. While he was out of the hospital, urine collections were provided, indicating a degree of conscious cooperation on the part of the patient, but the labeling was unreliable and the samples were useless (three were labeled 8 A.M. on the same day, for example). Data collection was more reliable in the hospital: "However, ——— returned to the V.A. Hosp. last week, and I believe collections began again last Wednesday, March 30th [1955]. So we should have better success from now on, I hope."[85] A year later, the sample collection problem persisted with this patient, who was surprisingly long-lived for one diagnosed as terminally ill.[86]

Diverging Goals

Deeper problems, which probably led to the end of the experiment, involved the basic design of the experiment and the types of data to be collected. As mentioned earlier, although the physicians and the health physicists had the same general interests, when it came to specifics, the two research programs had differing goals. As time passed, this problem increased.

First, the AEC was not getting the long-term autopsy data it wanted. Of the eight subjects autopsied for the study, seven died within four months of the injection.[87] Only one survived for seventeen months. Given Dr. Sweet's selection of terminal patients, this was not surprising, but the AEC had hoped for subjects who would survive at least eighteen months. Autopsies would then have provided tissue samples indicating the body's long-term deposition of uranium.

Secondly, Sweet's team wished to keep raising the dose until a useful concentration of uranium showed up in tumors. Bernard later recalled that the first patient was given a dose of 6 milligrams of uranium and died two days later. "They got brain tumor samples. There was very little uranium present, but Sweet was still wondering: maybe not a high enough dose. So we went as high as, I think, 50 milligrams in one patient, an intravenous injection in the anticubital vein."[88] The MGH team was also interested in when "toxic" effects would first appear as the doses increased. This would determine the maximum single dose that could be administered prior to neutron beam therapy.

The health physicists, by contrast, were more interested in the effects of a gradual, cumulative dose. Struxness had returned from Boston and wanted the experiment halted because the doses were too high.[89] The project was cancelled in 1957.

Oak Ridge's need for low dose data was now even greater. The Boston Project data on kidney damage indicated that the permissible concentrations of uranium in the air of production plants might have to be reduced by a factor of ten, leading to "several profound changes in the methods of using uranium in industry."[90] But were these kidney effects due to the total dose, or to its administration in a single injection? Would the effects be less if the same total dose had been administered gradually over a long time? In 1958 Karl Morgan, head of health physics at Oak Ridge, suggested to Sweet that a program of smaller doses might answer this question and still serve a therapeutic purpose: "Further cooperative work including single and multiple injections at a trace level may prove to be of mutual benefit."[91] This time, Morgan was prepared to offer a formal contract, but Sweet apparently never took Morgan up on the offer. As late as 1963, Morgan proposed a cooperative project with Argonne Cancer Research Hospital for studies on uranium excretion using low doses.[92]

Published Portrayal of the Human Subjects

The published reports' portrayal of the subjects exhibited two notable traits: first, the subjects were dehumanized, losing not only individuality but also consciousness and prospects for any recovery; and second, difficulties and errors were overlooked. Of course, neither of these traits is unusual in scientific reports of human studies.

The selection of terminal patients, in part to obtain autopsy data, occasionally led, understandably, to somewhat callous language in private correspondence: "Patient V is continuing downhill so you should be hearing from me soon."[93] This same tone continues in published reports. Their condition is described as "severe irreversible central nervous system disease. Virtually all had brain tumors of a most malignant type."[94] The term "virtually" avoids the difficult question of why the study was not limited to patients who actually had brain tumors; no mention is made of the one subject who, on autopsy, was found to be without tumor. Readers were also told that "at the time of injection all patients were in coma . . . receiving the usual hospital care consisting of frequent turning, skin care, gastric tube feedings, catheter drainage and frequent tracheal suction."[95] No mention is made of one patient who became ambulatory, left the hospital, and survived for over a year. The report only notes: "The patients who did not terminate during the two to three week period following injection were transferred to a nursing home where they could be closely observed."[96] The report published by health physicists describes all eleven subjects as "terminal brain tumor patients. . . . With the exception of 1 patient, who lived 17 months after injection, all were in coma or semi-coma."[97]

The rhetoric of scientific articles provides the authoritative voice of science and provides models for future researchers. It should not be confused, however, with the unfiltered daily reality of dealing with such patients. At the risk of being labeled an apologist, my own impression from interviewing both Sweet and Luessenhop, albeit some forty years after the event, is that they were sincerely committed to treating patients suffering from a tragic disease. But the sterile prose conflicts with the living experience of subjects and their families.

For example, as the study progressed, a further step was added of comparative biopsies from samples obtained during surgery.[98] These procedures may have been those which led to the most sensational of the accusations made some forty years later. In 1994, Lenore Fenn testified before ACHRE, recollecting what she had witnessed as a 17–year-old summer employee in 1955:

> A medical student invited me to "help" with procedures conducted by William Sweet, M.D. . . . The circumstances were unusual. We gathered in a remote part of the hospital after 10:00 P.M. Jacob Lefton, sedated and restrained, was on the operating table. A large rectangular flap was lifted from his cranium. Dr. Sweet placed probes connected to his brain tumor tissue, his normal brain tissue, his chest, thigh, leg, etcetera. He was then adminis-

tered a radioactive isotope, and as multiple scanners started to whiz, records were kept to indicate the relative concentration of radioactivity in various body parts. Brain and tumor tissue spilled onto the floor. Meanwhile, Jacob Lefton wept, struggled, prayed and cried out for help. When I expressed concern for his suffering, I was told that he would "not remember the pain as pain." . . . Two weeks later, we gathered again at night, and Jacob Lefton was treated with a different isotope. I learned that the point of the process was to discover which of several isotopes concentrated most intensely in Jacob's tumor tissue. I assumed that the concentrated radioactivity was healing Jacob's tumor. Everything went as before. Dr. Sweet was utterly absorbed in Jacob's brain and the scanner's numbers. There was a German doctor who went, although it is hard to imagine, right up to his elbows in Jacob's brain. Even more brain tissue and tumor tissue spilled onto the floor. Jacob wept, struggled, prayed and cried as before."[99]

Needless to say, such a vivid recollection was disturbing, but some details—such as a surgeon putting his arms up to the elbows into the patient's brain—weakened its credibility. Speaking off the record, one neurosurgeon active in the 1950s offered an explanation for what Ms. Fenn may have seen. At the time, a purely surgical therapy for brain tumor patients was to remove a flap of skull, cut out what tumorous tissue could be reached, and leave the flap of skull open. As the tumor grew, it would be "shaved" away. The therapy was especially gruesome, and since it did not extend the life span of the patients, it was eventually abandoned. This surgical therapy, combined with the uranium injection experiment, may be what Ms. Fenn recalled. This interpretation is supported by some of her own testimony. As noted above, she recalled that "a large rectangular flap was lifted from his cranium." Another possibility is that she saw the use of radioisotopes, such as phosphorous-32, to assist the surgeon in locating the boundaries of the brain tumor. Whatever the explanation, she recollected an experience quite different from that of a surgeon for whom brain surgery was a daily, if messy, activity.

The Experimental Results

Ironically, although the Boston Project centered on radioactivity, its main finding was that chemical, not radiological, effects were the primary danger from uranium exposure. The published reports described the primary purpose of the experiment as determining the single maximum dose that could be delivered without "toxic" effects. By toxic, the researchers meant only an observed, unfavorable change in vital signs, not a dramatic worsening of the patient's condition. According to one of the published reports, no "consistent or marked" changes were noted in blood pressure, pulse, respiration, body temperature, dermal changes (erythema, sweating, or eruptions), gastric residua, stools, or neurologic status. No notable changes were consistently seen in cardiac status, glucose tolerance, or liver function.[100] A drop in hemoglobin was observed in all the patients, but this was attributed to the loss of the 300 to 400 cc of blood withdrawn for testing.[101]

However, in the three patients who received the highest doses (aged 47, 63, and 39 years) analyses of urinary catalase and protein showed a double peaking: the first during the second to fourth day, and the second during the sixth to eighth day. The magnitude of the peak for albumin grew with increasing dose.[102] The explanation for such double peaking in animals was that the first peak represented a breakdown of membrane permeability in the tubular cells of the kidney, and the second peak the actual breakdown of cell structure. The appearance of casts in the urine of two patients was interpreted as further evidence of such breakdown.[103]

This double peak was interpreted as the first sign of a toxic chemical effect of uranium on the kidney. Thus, for uranium, the kidney was the "critical organ" determining toxicity—not bone, as with radium.[104] The mathematical models used by health physicists to calculate internal radiation dose were irrelevant so far as uranium toxicity was concerned: "The safe burden in the kidney, dictated by considerations of chemical toxicity, is one-tenth the burden deemed permissible from radiological considerations."[105]

The applicability of animal data was now questionable. The biological half-life of uranium in bone for humans, for example, was found to be one-fourth that of rats.[106] Analysis of blood and soft tissues showed "a significant fixation of uranium in soft tissues (other than kidney) not found in small animals."[107] A specific result was the modification of the elaborate mathematical model being developed for human metabolism of uranium.[108] But the results of the experiment apparently had no impact on existing radiation exposure limits. When compared with the older model, the new model yielded a maximum value for concentration of uranium in urine which was "a factor of 5 higher than the present value employed, thereby indicating a margin of safety for exposure to soluble compounds."[109] What is not mentioned is that this safety factor of five was a departure from the traditional safety factor of ten.

This human data from such a controlled exposure to uranium are so rare that they have been reanalyzed with the passage of time. As late as 1975 one researcher, Pat Durbin, sought to reexamine the data, "to obtain a more realistic description of the early behavior of uranium in human bone."[110]

Concerning Sweet's dream, the results were disappointing. The highest brain tumor count in the Boston Project was on patient IX, with 4 counts per minute; but the skull was higher at 4–13 counts per minute.[111] The data for concentration in brain tumor were quite scant compared with that for excretions and tissues.[112] The differences in concentration were not enough to make neutron beam therapy useful. The uranium program was never continued into a therapeutic stage, although there is evidence of later mouse experiments comparing the uranium concentration in brain and tumor.[113]

Conclusion

When we examine the Boston Project closely, we see a dramatic difference in the trust given to those responsible for weapons-grade uranium and those responsible for human subjects. Uranium allocation was reviewed at several levels. Those handling weapons-grade uranium were required to document every step of its custody and to account for it down to the milligram level. They were accompanied by armed guards when transporting it and carefully recorded its transfer to others.

In contrast, those with custody of the terminal human subjects were trusted as autonomous professionals. Animal experiments were not required first. Review and oversight of subject selection was superficial. No requirements were imposed regarding patient selection, patient consent, evaluation of risks, or pain versus benefit to the patients. Recently, the experiments have been the subject of a class action lawsuit.[114] As ACHRE stated in its final report: "That people are not likely to live long enough to be harmed does not justify failing to respect them as people."[115]

NOTES

Gilbert Whittemore's work on this chapter was supported in part by NSF grant #9529420.

1. Advisory Committee on Human Radiation Experiments, *Final Report* (Washington, D.C., 1995) (hereafter ACHRE-GPO); Advisory Committee on Human Radiation Experiments, *The Human Radiation Experiments: Final Report of the Advisory Committee on Human Radiation Experiments* (New York and Oxford, 1996) (reprint with index added) (hereafter ACHRE-OXF); Eileen Welsome, *The Plutonium Files: America's Secret Medical Experiments in the Cold War* (New York, 1999).

2. Four published reports resulted from the experiment: E. G. Struxness, A. J. Luessenhop, S. R. Bernard, and J. C. Gallimore, "The Distribution and Excretion of Hexavalent Uranium in Man," *Proceedings of the International Conference on the Peaceful Uses of Atomic Energy Held in Geneva 8 August–20 August 1955*, Vol. 10, *Radioactive Isotopes and Nuclear Radiation in Medicine* (New York, 1956), 186–96; S. R. Bernard, J. R. Muir, and G. W. Royster Jr., "The Distribution and Excretion of Uranium in Man," *Proceedings of the Health Physics Society* (June 1956): 33–48; A.J. Luessenhop, J. C. Gallimore, W. H. Sweet, E. G. Struxness, and J. Robinson, "The Toxicity in Man of Hexavalent Uranium Following Intravenous Administration," *American Journal of Roentgenology* 79 (Jan. 1958): 83–100; and S. R. Bernard, "Maximum Permissible Amounts of Natural Uranium in the Body, Air and Drinking Water Based on Human Experimental Data," *Health Physics* 1 (1958): 288–305.

3. Quoted in K. F. Eckerman, Martin Marietta Energy Systems, Inc., Oak Ridge, to Barry A. Berven, "The Boston-Oak Ridge Uranium Study," memorandum, 7 January 1994, Human Subjects Project: A-00001; DOE Bates # 1024032, Repository: Oak Ridge / Energy Systems / ORNL X-10, Collection: Health Sciences Research Div. / Keith F. Kerman Files, Bldg 7059, Correspondence. Copies of all other documents cited are also available in the archives of the Advisory Committee on Human Radiation Experiments held by the U. S. National Archives and Records Administration, Washington, D.C. (hereafter ACHRE Archives).

4. ACHRE-GPO (n. 1 above), 269, see also 262–64, 268; ACHRE-OXF (n. 1 above), 163, see also 158–60.

5. Interview of Dr. William Sweet by Gilbert Whittemore, ACHRE Staff, 8 April 1995, ACHRE Archives (hereafter, Sweet interview). Sweet died on 22 January 2001.

6. Sweet interview (n. 5 above).

7. Ibid., 1–2.

8. Ibid.

9. Ibid., 4–5.

10. Ibid., 6.

11. Ibid., 28.

12. The paper was written by Konger and Giles. Sweet interview (n. 5 above), 11.

13. Luessenhop (n. 2 above), 83.

14. Interview of Alfred Luessenhop, M.D., by Gilbert Whittemore, ACHRE staff, 8 March 1995, ACHRE Archives (hereafter Luessenhop interview), 20.

15. Sweet interview (n. 5 above), 17.

16. Sweet interview (n. 5 above), 19.

17. I. Farr, W. H. Sweet, J. S. Robertson, C. G. Foster, H. R. Locksley, D. I. Sutherland, M. L. Mendelsohn, and E. E. Stockton, "Neutron Capture Therapy with Boron in Treatment of Glioblastoma Multiforme," *American Journal of Roentgenology, Radiation Therapy and Nuclear Medicine* 71 (1954): 279–94; J. T. Godwin, L. E. Farr, W. H. Sweet, and J. S. Robertson, "Pathological Study of Eight Patients with Glioblastoma Multiforme Treated by Neutron Capture Therapy Using Boron-10," *Cancer* 8 (1955): 601–15.

18. Sweet interview (n. 5 above), 13.

19. Bernard, "Distribution and Excretion of Radium in Man" (n. 2 above), 33.

20. Ibid., 35.

21. Newell Stannard, transcript of interview with Dr. Bob Bernard, 17 April 1979, ACHRE Archives (hereafter Bernard interview), 3.

22. Bernard interview (n. 21 above), 4.

23. Luessenhop (n. 2 above), 84.

24. See Welsome (n. 1 above).

25. Bernard interview (n. 21 above), 4.

26. Ibid., 7.

27. Ibid.

28. Advisory Committee for Biology and Medicine, "Minutes Advisory Committee for Biology and Medicine, Thirty-eighth Meeting held at Cancer Research Institute, Boston, Massachusetts, June 26 and 27, 1953," U.S. DOE Archives, 326 U.S. Atomic Energy Commission, Division of Biology and Medicine, Box 1, Folder 8, 3.

29. Alfred J. Luessenhop, M.D., Department of Surgery, MGH, to Struxness, Oak Ridge Institute of Nuclear Studies, 28 August 1953; Human Studies Project D-00311; Repository MMES/X 10, Collection: Health Sci. Research Div., Box 1060 Commerce Park, folder: Room 253 (hereafter HSRD Box 1060).

30. Struxness (n. 2 above), 186.

31. Handwritten notes, apparently in response to a 1 February 1954 letter from Luessenhop to Gallimore. In the letter, Luessenhop had used the phrase "in the event of an autopsy." The handwritten note's comment: "1. 'In the event of an autopsy etc.' Is there a possibility that we will not perform an autopsy? This is bad! 2. Is it not understood that I will be present at an autopsy?" HSP D-00303, HSRD Box 1060.

32. Luessenhop to Gallimore, letter, 1 February 1954, HSP D-00339, HSRD Box 1060.

33. Gallimore to Sweet, 22 March 1954; NARA Atlanta Archives, Collection: RG 326 68A1096, "OR Research Division, Med, Health & Safety, etc.," Box 104, Folder: "Medicine, Health & Safety, Operation Boston."

34. Struxness (n. 2 above), 186.

35. Luessenhop (n. 2 above), 83.

36. Luessenhop interview (n. 14 above); Gallimore to Dr. John Schole [*sic*], Dept. of Neurosurgery, MGH, letter, 24 February 1955, HSP D-00355, HSRD Box 1060; Gallimore to Dr. John A. Scholl [*sic*], letter, 31 March 1955, HSP D-00359, HSRD Box 1060.

37. S. R. Bernard and E. G. Struxness, "A Study of the Distribution and Excretion of Uranium in Man: An Interim Report," ORNL Report No. 2304, Unclassified, 4 June 1957, DOE Bates # 1025923-1025980 (hereafter ORNL 2304), 3; Luessenhop to Struxness, 28 August 1953; HSP D-00311; HSRD Box 1060.

38. ORNL 2304, 4.

39. "As was decided at our conference, we will be able to supply you with blood and urine specimens as controls, then every 1 hour for 24 hours, then every 12 hours for 24 hours, then every 24 hours for as long a period as possible. All fecal specimens will also be included." Luessenhop to Struxness, 28 August 1953; HSP D-00311, HSRD Box 1060.

40. Luessenhop to Struxness, 28 August 1953 (n. 39 above). The "supplemental proposals" probably were additional biopsies that were conducted on bone, muscle, and blood, in addition to cerebrospinal fluid, in the 60 hours following injection. See detailed summary of sample collection and analysis in Gallimore to Sweet, 22 March 1954, NARA Atlanta Archives, Collection: RG 326 68A1096, "OR Research Division, Med., Health & Safety, etc.," Box 104, Folder: "Medicine, Health & Safety, Operation Boston."

41. "Temporarily, the only uranium determinations, other than blood, urine and feces, which we will need will be on cerebralspinal fluid." Luessenhop to Struxness, 17 September 1953; Human Studies Project D-00312, HSRD Box 1060.

42. Struxness, (n. 2 above), 186.

43. "With increasing dosage . . . " Luessenhop (n. 2 above), 97

44. ORNL 2304, 5. Unsigned, undated handwritten notes entitled "Project Boston" listing tasks, the first of which is "1) Plutonium, a) set up operations (injection, sample prep., autopsies) at Brookhaven, b) make arrangements with BNL HP to provide all services." It is unclear whether these were tasks already underway or speculation about possible future research. HSP D-00325, HSRD Box 1060.

45. ORNL 2304, 24.

46. Struxness to F.C. Uffelman, Bldg. 9706-1A, "Request for SP Material," 28 September 1953; HSP D-00313, HSRD Box 1060. Struxness, Individual Requestor, "Request for Research Issuance," 12 January 1954. By January 1954 the request had risen from the initial 13 Mg of uranium-233 to 50 Mg of uranium-233 and 500 Mg of uranium-235. Unsigned handwritten notes, "U-238, U-233 solution for Boston Project," HSP D-00318, HSRD Box 1060.

47. Struxness to Uffelman, (n. 46 above).

48. L. B. Emlet, Y-12 Plant Superintendent, Carbide and Carbon Chemicals Company, to Mr. E. C. Armstrong, Atomic Energy Commission, Oak Ridge, "Request for Diversion: 12.5 Mg. U-233," 5 October1953; HSP D-00316, HSRD Box 1060.

49. Unsigned handwritten notes (n. 46 above).

50. Ibid.

51. Unsigned memorandum from Health Physics Department to Mr. F. C. Uffelman, Carbide and Carbon Chemicals Company, Oak Ridge, "Issuance and Transfer of Uranium to Dr. Sweet," 7 December 1953; HSP D-00333, HSRD Box 1060.

52. H. C. Armstrong, Production Division, Atomic Energy Commission to Mr.

L. E. Emlet, Carbide and Carbon Chemicals Company, 3 November 1953; HSP D-00321, HSRD Box 1060.

53. Description of solutions shipped with handwritten notes confirming receipt, 31 October 1953, HSP D-00319, HSRD Box 1060.

54. Struxness to Uffelman, "Request for Research Issuance of SF Material," 22 December 1953, HSP D-00337, HSRD Box 1060.

55. Luessenhop to Struxness, 28 August 1953; HSP D-00311, HSRD Box 1060.

56. "Application for Approval of Use of Radioactive Isotopes, Massachusetts General Hospital," submitted by Dr. William Sweet on 30 September 1953, approved 3 November 1953, ACHRE Archives (hereafter Sweet application).

57. Sweet application (n. 56 above).

58. Ibid.

59. Ibid.

60. Ibid.

61. Bernard interview (n. 21 above), 8.

62. Luessenhop to Struxness, 3 November 1953, HSP D-00320, HSRD Box 1060.

63. Ibid.

64. Luessenhop to Struxness, telegram, 18 November 1953; HSP D-00323, HSRD Box 1060.

65. Luessenhop to Struxness, 30 November 1953; HSP D-00331, HSRD Box 1060.

66. Janette Robinson, Res. Asst. to Dr. Wm. H. Sweet, to John Gallimore, Health Physics Dept., Plant Y-12, Oak Ridge, letter, 6 April 1955; HSP D-00360, HSRD Box 1060.

67. Luessenhop to Gallimore, 1 May 1954; HSP D-00342, HSRD Box 1060; Robinson to Gallimore, 3 February 1955; HSP D-00351, HSRD Box 1060.

68. Robinson (n. 66 above).

69. Sweet interview (n. 5 above), 21.

70. Robinson (n. 66 above).

71. Luessenhop (n. 2 above), 93.

72. Sweet interview (n. 5 above), 20.

73. Ibid., 21.

74. ACHRE, Transcript of Proceedings, December 19, 1994, ACHRE Archives, 2ff.

75. Ibid., 2ff.

76. ACHRE-GPO (n. 1 above), 263; ACHRE-OXF (n. 1 above), 159.

77. Struxness to Uffelman, "Accountability of First Shipment of Uranium to Mass. General Hospital," 21 December 1953; HSP D-00336, HSRD Box 1060.

78. Luessenhop (n. 2 above), 85.

79. E. F. Bernard (?), Typed notes entitled "Boston Project," 25 February 1954, HSP D-00310, HSRD Box 1060.

80. Struxness (n. 2 above), 187.

81. Gallimore to Sweet, 22 March 1954, NARA Atlanta Archives, Collection: RG 326 68A1096, "OR Research Division, Med., Health & Safety, etc.," Box 104, Folder: "Medicine, Health & Safety, Operation Boston."

82. Luessenhop to Gallimore, 1 February 1954, HSP D-00339, HSRD Box 1060.

83. Handwritten notes, "Telephone communication between Gallimore, Sweet & Struxness, November 2, 1954," HSP D-00348, HSRD Box 1060.

84. "He is now at Holy Ghost Hosp. and we are not collecting urine any longer." Robinson to John [Gallimore, Oak Ridge], handwritten letter, 14 March 1955, HSP D-00358, HSRD Box 1060.

85. Robinson (n. 66 above).

86. Robinson to Bernard, letter, 6 April 1956, HSP D-00368, HSRD Box 1060.

87. Bernard (?) (n. 79 above).

88. Bernard interview (n. 21 above), 8.

89. Interview of Karl Morgan by Gilbert Whittemore and Miriam Bowling, ACHRE staff, 6 January 1995, ACHRE Archives. Morgan died on 8 June 1999.

90. Karl Z. Morgan, Director, Health Physics Division, Oak Ridge, to Sweet, letter, 16 July 1958, ACHRE Archives.

91. Ibid.

92. Morgan to W. H. Jordan, Inter-Laboratory Correspondence, Oak Ridge, "Proposed Study of Distribution and Excretion of Enriched Uranium Administered to Man," memo, 2 September 1963; HSP A-00125; Depository: Oak Ridge / Energy Systems / ORNL (X-10), Collection: Office of Radiation Protection, Box: K. Z. Morgan's Files (Past Div. Dir. of Health Physics); Folder: Bld. 4500 s, Wing 4, Attic.

93. Luessenhop to Gallimore, 1 May 1954; HSP D-00342, HSRD Box 1060.

94. Struxness (n. 2 above), 186. This study reports only on patients aged 26, 34, 39, 47, 60, and 63 years old, which correspond to the patients numbered I, III, V, II, VI, and IV, who were injected with hexavalent uranium.

95. Struxness (n. 2 above), 186.

96. Ibid.

97. Bernard, "Distribution and Excretion of Radium in Man," (n. 2 above), 35.

98. Struxness (n. 54 above).

99. ACHRE, Transcript of Proceedings, 19 December 1994, ACHRE Archives, 2ff.

100. Luessenhop (n. 2 above), 85.

101. Ibid., 87

102. Ibid., 97.

103. Ibid., 98.

104. ORNL 2304, 25.

105. Bernard, "Maximum Permissible Amounts of Natural Uranium" (n. 2 above), 289.

106. Struxness (n. 2 above), 193. The authors later note that the condition of the patients, especially a negative calcium metabolism, would "hasten the removal of uranium from the skeleton." Bernard, "Maximum Permissible Amounts of Natural Uranium" (n. 2 above), 195.

107. Bernard, "Maximum Permissible Amounts of Natural Uranium" (n. 2 above), 195.

108. For details, see Bernard, "Distribution and Excretion of Radium in Man," (n. 2 above), 40ff.

109. Ibid., 46.

110. Patricia W. Durbin, Lawrence Berkeley Laboratory, to John Poston, Health Physics Division, Oak Ridge National Laboratory, letter, 21 November 1975; HSP D-00370, HSRD Box 1060.

111. Luessenhop to Bernard, telegram, 20 March 1956; HSP D-00367, HSRD Box 1060.

112. ORNL 2304, Table VI, p. 32. I have not found any journal article using this data on the concentration of uranium in the tumors.

113. "Contents of shipment: Blood Specimens from mouse experiment," 10 June 1954 and "Mouse specimens," 29 July 1954; HSP D-00344, HSRD Box 1060.

114. The history of a 1997 class action lawsuit filed on behalf of subjects of boron-neutron experiments (*Heinrich et al vs. Sweet et al*, Civil Action No. 97-CIV-12134-MLW, United States District Court, District of Massachusetts) is beyond the scope of this paper.

115. ACHRE-GPO (n. 1 above), 269; ACHRE-OXF (n. 1 above), 163.

CHAPTER EIGHT

Writing Willowbrook, Reading Willowbrook

The Recounting of a Medical Experiment

Joel D. Howell and Rodney A. Hayward

In the history of human experimentation, a number of often-cited examples stand out. Some of those examples have been the object of attention as models for good; others, more often, for evil. Some experiments have been in the public eye only briefly; others have remained controversial for decades. In the latter instance, the stories of the experiment have been created and re-created, time and time again. Examining how and why the various stories were created reveals a great deal not only about the experiment and experimenters but also about those who (for a wide range of reasons) have chosen to study the events in question. This chapter will consider a series of experiments that remain controversial to this very day—a series for which there remains loyal supporters as well as committed critics. We hope that by obtaining insight from both the experiments and the ways that they have been interpreted (and reinterpreted) we may gain insight into the ways that historians ought to consider human experimentation.

For more than three decades the stories of the hepatitis studies begun in the 1950s at the Willowbrook State School have been told and retold, interpreted and reinterpreted. In so doing, the stories of these experiments have been used as the starting point for intense, heated, and occasionally insightful debate. Based on feeding children live virus in a controlled clinical setting, these experiments described differences between two types of viral infections of the liver, infections that are now called hepatitis A and hepatitis B. When the studies were first reported at pediatrics meetings, hepatitis symposia, and, soon thereafter, on the august pages of the *New England Journal of Medicine*, most listeners appear to have heard simply a tale of good science, of elegant experiments, of concerned pediatricians making important discoveries about the basic nature of hepatitis for the benefit of humankind. Yes, the experimenters transferred the infectious agent from child to child using fecal extracts, but the feces was prepared with great care. It was centrifuged, heated, and treated with penicillin and chloramphenicol, two of the most powerful antibacterial drugs known at the time. Before using the stool extract to pass the disease on to children, the investigators established its safety by inoculating first tissue culture preparations, then forty-seven mice, and finally five monkeys. Even after all this preliminary testing, when the experimenters began feeding the feces to children (mixed with chocolate milk) they started rather "gingerly," at a 1:100,000 dilution of stool, and only very slowly worked their way up to a 1:5 dilution.[1]

The fact that this tale was fundamentally based on feeding virus-laden stool to "mentally retarded," indigent children on the wards of a state mental hospital at first elicited little critical comment.[2] Perhaps this was because the authors explicitly noted that the "decision to feed hepatitis virus to patients at Willowbrook was not taken lightly," and they specifically outlined seven justifications for their actions.[3] Perhaps the lack of critical reaction was because most of the listeners were physicians who saw the Willowbrook experiments as consistent with, and logical extensions of, the enormous amount of previous human volunteer experimentation on infectious hepatitis—a disease thought to be caused by a virus, but a virus that was at that time not possible to culture (and thus to study) except as it could be made to grow inside of human beings.[4] But perhaps most important, listeners probably were not concerned because they believed the authors' claims that the appallingly bad conditions at the Willowbrook State School meant that all of the children were destined to become infected anyway, with or without the experimenters' interventions. Thus, it mattered little whether the children acquired hepatitis "naturally" or through the active intervention of physicians; they would eventually become infected with hepatitis, one way or another.[5]

In this chapter we will focus on this claim that hepatitis infection was an inevitable result of hospitalization at the Willowbrook State School. We will consider both how the claim was written by investigators and how others have read the claim, by the many people who have supported and the many others who have criticized the Willowbrook experiments. And critics there have been, ever since the mid-1960s when a Harvard anesthesiologist named Henry K. Beecher chose to include the Willowbrook experiments as the supposedly disguised but in fact quite easily recognizable "Case number 16" of his now-famous article on ethics and clinical research.[6] Subsequent attacks on the ethics of the experiments have come from a strikingly wide range of perspectives. The debate has been played out with a passionate intensity and a virulent personal quality rarely seen within the normally subdued halls of academe. Starting with the obvious affront to aesthetic sensibilities evoked by the image (and the reality) of feeding feces to children, criticisms have centered on such questions as whether children should ever be used for experimentation, whether mentally retarded children should be used for such purposes, whether the consent given by parents was coerced, and whether the experiments were done in accordance with broad ethical codes or with the approval of the appropriate oversight committees.

However, unlike many other controversial medical experiments, mention of the Willowbrook studies awakens a core of committed people in high places who come rapidly and enthusiastically to their defense. Supporters have attempted to rebut every element of the critics' concerns.[7] To consider carefully all of these aspects of the debate would lead to a much longer work and would take us far afield from the fundamental issue we wish to analyze here, which is when and how historians ought to engage with past incidents of human experimentation.

We shall address this question by considering in some detail one major justification that has been offered from the very beginning: that the disease was inevitable, that hepatitis "affected virtually every child at Willowbrook."[8] This assertion seems fundamental to any discussion of Willowbrook. Were this claim believed not to be true, there would be little to debate; indeed, the experiments probably would not have been done. Within the heated discussion about Willowbrook, supporters and critics alike have nearly universally accepted the claim of inevitable infection. For example, when in 1970 the philosopher Paul Ramsey considered the hepatitis experiments, he disapproved of much that went on but simply quoted the original accounts that "it was apparent that most of the patients at Willowbrook were naturally exposed" without questioning to whom it was apparent—or, more to the point, precisely when it became apparent.[9] It behooves scholars to problematize the issue, to consider both the evidence for that claim and the readings of that evidence.

On the surface, the assumption of disease inevitability seems reasonable. There can be little doubt that the Willowbrook State School in the mid-1950s was a horrific place, later to become the subject for media exposés and judicial debates. But in the 1950s, when a team of physicians from New York University Department of Pediatrics entered the institution, Willowbrook was not the site of an overtly public and political controversy. It was simply a dismal, overcrowded, natural breeding ground for all manner of infectious diseases.[10]

Based on their initial observations, the experimental team from New York University claimed that virtually all of the children admitted to Willowbrook were destined to become infected with hepatitis. One could consider this claim in terms of at least three different sets of interrelated questions. First, what did the investigators themselves believe to be true when they started the experiments in the mid 1950s? Second, what was the published data on which their claim rested, and how has that data been read? And, third, what do we now believe to be true about the inevitability of disease at Willowbrook in the 1950s? Absent the ability to talk to investigators at the time that they were setting up the experiments, the answer to the first question is probably unanswerable.[11] The answer to the third question rests on techniques that were unavailable to investigators in 1950s and is thus not germane to historical questions about the actions of the investigators. We shall focus on the second question, the claims investigators made in the published literature about the inevitability of children acquiring hepatitis. The evidence to support this claim was not obscure; it was published in major journals in tabular form with the very earliest experimental reports. After examining this evidence, we shall consider the response of critics and historians, and of policy analysts, looking at the discussion as it was carried out in the public eye.[12] Before doing so, however, it will be useful to consider the conceptual environment within which the Willowbrook experimenters started their work.

Infectious Hepatitis: Prospects and Problems in the 1950s.

Around the start of World War II it was generally believed that there were two types of hepatitis. One type, infectious hepatitis, easily passed from person to person through an oral-fecal route, often occurred in epidemics, ordinarily had no serious sequelae, and was usually milder in children than in adults. Another type, serum hepatitis, was contracted via either blood or blood products, or through maternal-fetal transmission. Infectious hepatitis had been described, literally, since antiquity.[13] Serum hepatitis had been named in the late nineteenth century, but it became a topic of increasing attention with the increasing use of blood and blood products in the 1940s and 1950s.

During World War II both types of hepatitis posed major problems for military forces.[14] After the war, infectious hepatitis continued to be a major problem for institutionalized populations. Gamma globulin provided some protection, but its effects were temporary.[15] In order to control the spread of infectious hepatitis, physicians needed to know how long people with the disease remained infectious. Because the causative virus could not be cultured, the only way to assay feces for the presence of hepatitis was by using human volunteers. If the volunteer ingested feces and then contracted hepatitis, that was evidence that the feces had come from a patient who was still capable of passing on the disease. This approach limited investigators' abilities to study hepatitis. In one study the investigators apologized: "The number of subjects employed in testing various materials is necessarily small due to the difficulty in obtaining human volunteers."[16] Indeed.

Diagnosing infectious hepatitis was problematic. Jaundice was a characteristic symptom of liver disease. But investigators realized that liver disease such as hepatitis need not always produce jaundice—that is, disease could be non-icteric (also referred to as anicteric).[17] In some instances, hepatitis might not produce any symptoms at all. Or hepatitis might cause only nonspecific symptoms such as nausea and fatigue. Thus, people with anicteric or asymptomatic disease might not realize that they had hepatitis, or caregivers for an institutionalized person might not realize that the person was sick. Asymptomatic or anicteric people with hepatitis who did not take appropriate precautions to slow the spread of the disease could be particularly important vectors for disease transmission.

Two other means of detecting disease were of increasing interest. A method for obtaining liver tissue through a biopsy instrument passed through the skin had recently become available.[18] This new technology enabled correlation of the microscopic, histological features of the diseased liver with clinical findings. Moreover, in the early 1950s a series of new blood tests were being described that could potentially identify patients with liver disease.[19] Such tests were held to be particularly useful if they could not only diagnose patients with disease but could also differentiate "surgical jaundice," caused by diseases such as gallstones that would require surgical intervention, from "medical jaundice," caused by diseases such as infectious hepatitis for which surgical intervention would be useless (and perhaps dangerous).[20] Thus, the study of hepatitis was an obvious place for a bright, young clinical investigator to turn in the 1950s. It was an exciting disease, an important disease. And although it posed a set of complex problems, there also seemed to be an abundance of new ideas and tools that might be used to solve those problems. It was in this context that pediatricians viewed the problem of hepatitis at the Willowbrook State School.

Writing Willowbrook: Claims Made by Investigators about the Incidence of Disease

The starting point for reading claims about the incidence of hepatitis at the Willowbrook State School is the information presented by the investigators. Saul Krugman, a pediatrician at New York University Medical School who lead the Willowbrook experiments for several decades, was first asked to consult at the school in 1954, where he noted a remarkably high incidence of infectious hepatitis. Given the extremely poor sanitary conditions, this finding was biologically plausible and was consistent with the well-known problem of hepatitis at such custodial institutions.

One of the first times Krugman presented his data about the incidence of hepatitis at Willowbrook was at the 1956 international symposium on hepatitis (table 8.1). Krugman and his associates claimed that in 1955 there were 106 cases of infectious hepatitis in Willowbrook patients, a yearly attack rate for patients of 25 per 1,000. Also of concern was the presence of 23 cases in a much smaller number of attendants, producing an attack rate for attendants of 40 per 1,000. Two years later, in 1958, Krugman and his collaborators published their first major paper on the experiments at the Willowbrook State School in the *New England Journal of Medicine* (table 8.2).

Table 8.1. *Age Distribution of Inmates and Cases of Jaundice, 1953–55*

| Age (years) | Inmates | | Jaundice | | |
	No.	%	No.	%	Rate/1000
0–4	535	12.8	16	8.6	30
5–9	1017	24.3	42	22.6	41
10–14	858	20.6	34	18.2	40
15–19	576	13.9	47	25.3	81
20–29	621	14.8	22	11.8	35
30–39	313	7.5	13	7.0	40
40–49	161	3.8	7	3.8	43
50+	96	2.3	5	2.7	52
Totals	4177	100.0	186	100.0	

(Inmates % for ages 0–4 through 15–19 bracketed together: 71.6)

Source: Reprinted from Robert Ward, Saul Krugman, Joan P. Giles, and A. Milton Jacobs, "Endemic Viral Hepatitis at an Institution: Epidemiology and Control," in Frank W. Hartman, Gerald A. LoGrippo, John G. Mateer, and James Barron, eds., *Hepatitis Frontiers* (A Henry Ford Hospital International Symposium) (Boston, 1957).

The two tables are clearly related. In one sense they represent different stages of a single work-in-progress. The earlier version of the table includes 186 cases; the later, more widely read paper in the *New England Journal of Medicine* includes 284 cases. But there is another difference: the 1956 table lists "cases of jaundice," while by 1958 the listing had been changed to "patients with hepatitis," although the text refers to "cases of jaundice."[21] The latter table has been used for several decades as the centerpiece for a principal justification of the Willowbrook studies.

Yet the tables appear at first glance not to substantiate the authors' textual claims. Consider the hepatitis attack rate for patients admitted to Willowbrook. The maximum attack rate for any age group in these two tables was roughly the same, 20 to 25 cases per 1,000 people per year, and the authors chose to emphasize that maximum value both in their 1958 publication and when they reported later on the results of additional experiments.[22] Over the ensuing years, the authors frequently referred to this data to justify their claims that the experiments were ethical. They progressed from the initial observation that "it was apparent that most of the patients at Willowbrook were naturally exposed to hepatitis virus"; went on to say that "it was inevitable that most of the newly admitted susceptible mentally retarded children would acquire the infection"; and contin-

Table 8.2. *Age Distribution of Inmates (Exposed Population) and of Patients with Hepatitis, 1953–56*

Age (years)	Inmates No.	Inmates %	Patients with Hepatitis No.	Patients with Hepatitis %	Age-Specific Attack Rate*
0–4	535	12.8	27	9.5	51
5–9	1017	24.3	68	24	67
10–14	858	20.6	56	19.7	65
15–19	576	13.9	58	20.4	100
20–29	621	14.8	34	12	55
30–39	313	7.5	17	6	54
40–49	161	3.8	14	4.9	87
50+	96	2.3	10	3.5	104
Totals	4177	100.0	284	100.0	

(The 0–4 through 15–19 inmate percentages are braced to 71.6; the corresponding hepatitis percentages are braced to 73.6.)

Source: Reprinted from Robert Ward, Saul Krugman, Joan P. Giles, A. Milton Jacobs, and Oscar Bodansky, "Infectious Hepatitis: Studies of Its Natural History and Prevention," *New England Journal of Medicine* 258 (1958): 407–16.
*Attack rate/1000 during 4-year period, 1953–56.

ued, using statements like "most children admitted to Willowbrook inevitably acquired hepatitis," "it was inevitable that susceptible children would become infected in the institution," and "hepatitis . . . affected virtually every child at Willowbrook."[23] It would seem, on the face of it, that an attack rate that is no higher for *any* age group than 25 cases per 1,000 per year hardly leads to the conclusion that it is "inevitable" that every child will get the disease. Yet these assertions have stood virtually unchallenged throughout over three decades of contentious discussion.

Why? Is there any way to understand how readers could make those numbers support a claim of certain disease? To attempt to do so, we must examine the 1958 table within the context of the times in which it was created and consider a number of assumptions that were largely unaddressed by the Willowbrook investigators.

First, we must ask if age-specific attack rates were assumed to be stable. The cumulative odds of acquiring hepatitis for the population as a whole would be highest if one assumed that children admitted to Willowbrook remained at the same age-specific risk for acquiring infectious hepatitis over time. One might now argue that this is probably not a valid assumption. As a result of behavioral or biological differences, some of the children admitted to Willowbrook would almost certainly have been more likely to acquire the disease than others. Those children who were at highest risk would have contracted the disease first, leaving uninfected many of those children at lower risk. Because the children most likely to become ill would contract the disease earlier, those less likely to acquire the disease would remain uninfected. The observed attack rates would thus be expected to decrease over time. It was known by the 1950s that epidemics tend to have a characteristic pattern, with a rapid rise of disease followed by a gradual tapering. In fact, the investigators themselves describe such a pattern, stating that the epidemic of hepatitis was new, with the attack rates rising between 1953 and 1955 and then falling for the next two years.[24] Later the investigators came to believe that the epidemic was different from epidemics at other institutions, that it was less explosive and more longlasting.[25] However, it is possible that the investigators believed that everyone at Willowbrook was at equal risk, and it is possible that the investigators had a hunch at the outset of their studies that the epidemic would continue unchanged for some time (although their data would suggest otherwise). Not having evidence to suggest any other interpretation, in order to increase the apparent cumulative likelihood of children at Willowbrook acquiring hepatitis we shall assume that the investigators believed that the epidemic would persist at the rate reported between 1953 and 1956.

The general belief was that a single virus caused infectious hepatitis and that people who were infected usually had long-term immunity to reinfection.[26] We do not have precise data on the age distribution of people admitted to Willowbrook, so we will assume that exposure would start at birth and that children remained at Willowbrook. If children were admitted later in their life, they would have less time to be exposed to hepatitis. By assuming that they will be exposed to disease from birth, we assume conditions that will predict the maximum cumulative probability of disease. We also assume that the risk varies by age as Krugman reported.

The final assumption we must consider has to do with whether all of the children who became infected with hepatitis would have been *diagnosed* as having hepatitis. Estimating precisely how many cases of hepatitis would likely have been anicteric or asymptomatic and thus might be unrecognized is critical for understanding how one might move from the numbers in the table to assuming that disease would be inevitable: not how many cases of hepatitis we would now believe would be unrecognized or how many the investigators later came to believe were unrecognized, but how many cases the investigators *at the time* believed would escape notice.

Here we have contemporaneous evidence of current thinking. Saul Krugman published widely on the subject of infectious diseases of children in general and on hepatitis in particular, and he specifically addressed the issue of anicteric hepatitis. His textbook on infectious diseases of children states that for infectious hepatitis, "The ratio of non-icteric to icteric cases is not known. It has been estimated in adults to be at least 1:1. In children it is probably greater."[27] A few years later one of the investigators cited an 8:5 ratio of anicteric to icteric cases.[28] In the 1958 *New England Journal* report, the authors state that "in all probability cases of hepatitis without jaundice were occurring with a frequency equal to overt forms" (i.e., a 1:1 ratio).[29]

But not all anicteric cases would have been asymptomatic. Anicteric cases might well have been symptomatic, albeit without jaundice as one of the symptoms. Krugman was well aware that one could at times make a diagnosis of infectious hepatitis without jaundice. In his textbook he discussed studies on manifestations of hepatitis in infants, many of whom did not have overt jaundice but did have nonspecific symptoms such as light-colored stools, fever, and vomiting.[30] Most patients at Willowbrook were not infants, although the precise age range was not published with the Willowbrook data.[31] The essential point remains, however, that some anicteric patients with hepatitis were known to manifest the disease in other ways. Thus, one might postulate that at least some of the patients with anicteric disease could still be diagnosed as having infectious hepatitis.

In fact, this was the case for some of the patients at Willowbrook. The change in designation on the tables from "cases of jaundice" in 1956 to "patients with hepatitis" in 1958 suggests that at least some patients included in the later table were recognized to have hepatitis without jaundice. In their earliest reports, the authors stated: "Doubtless there were many unrecognized examples of hepatitis without jaundice."[32] But with careful observation, Krugman believed that experimenters could lower the proportion of undetected cases. He stated that "in the majority of cases the diagnosis *would have been missed* if the patient had not been followed carefully with serial liver function tests and daily physical examinations" (emphasis added).[33] Thus, the numbers listed as "patients with hepatitis" included some anicteric and asymptomatic cases.

We shall analyze the tabular data in two ways. First, we shall assume that

Table 8.3. Expected Number of Residents Who Would Develop Hepatitis by Age Specified (cumulative incidence)

Age (years)	Assuming a 1:1 Using Published Data (%)	Ratio of Diagnosed to Undiagnosed Cases (%)
0	0	0
4	5.1	10.2
9	13.1	25.3
14	20.1	37.4
19	30.0	53.0

the experimenters' careful observations allowed them to detect all of the children who became infected with hepatitis, that the authors changed their title from "cases of jaundice" in 1956 to "patients with hepatitis" in 1958 because their "serial liver function and daily physical examinations" allowed them to diagnose all of the children who became infected. But in order to maximize the predicted incidence of disease that one could infer from the tables, we shall also analyze the data as if the only cases being reported were those children who became jaundiced. We shall assume for this second analysis that the investigators would have missed half the cases of hepatitis, the same ratio stated in the text of the article by the experimenters. This has the net effect of doubling the apparent incidence of disease.

Some of the above assumptions would clearly be labeled as "wrong" in the 1990s. Nonetheless, each would have been reasonable in the 1950s, and each would have the effect of increasing the perceived likelihood of the Willowbrook children acquiring the disease. We present the data using the same age groups as the experimenters (table 8.3).[34]

Thus, even after nineteen years of exposure—from birth to adulthood—only 30 percent of the children at Willowbrook would have been expected to be infected with hepatitis, based on the published data and generous assumptions. That number reaches only 53 percent if one assumes that half the cases were not being diagnosed. This level of probability seems far from the inevitability asserted in the many texts.[35]

New Data, New Estimates

Around 1960, based (in part) on additional experiments using the children at Willowbrook, investigators came to suspect that far more cases were going undiagnosed than they had previously estimated. They suggested a higher ratio of asymptomatic to symptomatic cases—a ratio of about 12:1.[36] If one revisits the attack rates from earlier studies using this new estimate, the likelihood of contracting hepatitis soon reaches the "inevitability" claimed by the investigators.[37] This revised estimate, of course, cannot be used to defend actions taken before the new data had been created.

By the late 1970s, thinking about hepatitis had changed a great deal. Thanks in large part to the Willowbrook studies, physicians now described two types of hepatitis, A and B. New serological techniques could reveal infection with hepatitis far more accurately than before and could easily differentiate between hepatitis A and hepatitis B. After residents had been at the school for three years, there was evidence for infection with hepatitis A in 97 percent of residents, a finding consistent with the revised, higher ratio of asymptomatic to symptomatic cases.[38] This data, investigators felt, was clear evidence of the validity of their earlier statements: "Our epidemiological surveys in the late 1950s indicated that 'it was inevitable that most newly admitted susceptible mentally retarded children would acquire the infection in the institution at large. This prediction was based on the observation of high attack rates of icteric and anicteric hepatitis infection.' Thus, these serological data confirm our earlier prediction that it was inevitable that newly admitted children to this high endemic area would acquire type A and type B hepatitis."[39] However, to make this argument, the authors cite their work from 1958 and 1959 in the *New England Journal of Medicine*, work that preceded any evidence that would substantiate such a claim, and work that cited a much lower, 1:1 ratio of icteric to anicteric cases. In order to evaluate the evidence that the Willowbrook investigators used to justify the decision to initiate the studies, we can use only the evidence that existed at that time.

Is there any reason not to take the investigators' quantitative assertions as they appear? The Willowbrook group published their results widely. After the revisions from the preliminary data in Table 1 (1956) to the later data in Table 2 (1958), they almost always used the same unqualified set of numbers. The investigators were fairly sophisticated and specific in their quantitative analyses. In one study they emphasized that they not only made the comparisons using a chi-squared test, but that they corrected that test for small numbers.[40] Their early articles include references to significance testing (which at times failed to reveal a statistically significant difference), suggesting some measure of quantitative sophistication on the part of the authors.[41] Thus, by the standards of clinical investigators of the time, the experimenters were capable of being rather precise in their presentation of data and estimation of probabilities.

One more assumption could make the numbers in these tables come closer to agreeing with the assertions of disease inevitability. If one believes that some of the children in each age group had already acquired the disease before entering Willowbrook, then the attack rate published in the paper for *all* children, including those who were already immune, would be lower than the true attack rate for a smaller group of *susceptible* children. Indeed, Krugman made this claim in a 1989 discussion with the author of an unpublished paper.[42] Krugman stated that the investigators felt that some of the children were immune because of earlier exposure to the disease. Thus, the number given as the denominator in the tables, 1,000, was a formal and artificial number. We note this claim, but we find this post-hoc reading to be uncompelling for at least three reasons. First, the authors of the Willowbrook studies state clearly in their earliest papers that children were acquiring the disease as they entered the institution; they refer to a "new outbreak" of infectious hepatitis.[43] Second, it seems unlikely that the investigators would make this assumption, yet not indicate anywhere in their published work that they were doing so. Such an assumption would imply that the reader should not take seriously the attack rates as given in the tables—which would have been substantial underestimates. Were that the case, one would have expected the experimenters to say so earlier and more publicly, for the experimenters—along with their critics—have certainly felt free to revise, expand, and clarify almost every part of the record. Finally, our historical understanding of the ethical assumptions that underlie these experiments should be based primarily on evidence that was generated at the time. Medical history in general, and the history of these experiments in particular, is replete with examples of retrospective clarifications. Thus, we should examine what experimenters said and wrote at the time of the events under consideration and note, cautiously and critically, post-hoc explanations offered many years later. Even if we provisionally accept this post-hoc explanation for purposes of analysis only, it remains hard to arrive at the notion of disease being "inevitable."

For example, if one assumes that 10 percent of children had acquired the disease before coming to Willowbrook and that half the cases were unrecognized, the 19–year cumulative incidence would only increase from 53 to 57.3 percent.

Reading Willowbrook

In our analysis we have consistently made assumptions in directions that would *increase* the estimated likelihood of children acquiring hepatitis and thus make it easier to substantiate the investigators' claims that disease was inevitable. But there is no reason to believe that critics of the experiments have felt a need to be nearly so generous. Yet the assertion that children at Willowbrook were going to become infected with hepatitis with or without the experimenters' interventions has gone curiously unquestioned.

Implicitly accepted by critics, the alleged inevitability of disease has been a cornerstone for the experiments' defenders, who over and over again have returned to this "fact." They have said that "these investigators have repeatedly explained—for over a decade—that natural hepatitis infection occurs sooner or later in virtually 100% of the patients admitted to Willowbrook"[44] and that "even without the Willowbrook studies, the children still would have become infected in due course by natural spread of the diseases in the institution."[45] These statements have been made in the face of widely distributed tables that are far from confirming such dire predictions. Questioning whether experimenters knew that disease was inevitable at the inception of the studies would undercut what defenders have clearly taken to be an essential element of their justification.

Why have the assertions about the inevitability of infection with hepatitis remained almost completely unchallenged?[46] Exploring how Willowbrook has been read may help us understand not only the Willowbrook experiments per se but also how other human experiments have been considered.

Numbers and Words. The lack of attention may partly result from how people use language to describe quantitative ideas. Although the Willowbrook researchers gave seemingly clear and precise numbers in their widely published tables, the debates were carried on using terms such as "most," "inevitably," and "virtually every." We cannot say precisely what the primary actors interpreted those words to mean or what their audience (critics or supporters) heard. There appear to be no studies from the period that evaluated what people meant by a particular verbal specification of the frequency of an event. But studies from the more recent past have asked both physicians and nonphysicians about the meaning of terms such as "always," "likely," and "certain." There are no generally shared beliefs about the meaning of commonly used verbal expressions of probability. For example, only 80.3 percent of people think that "certain" means 100 of 100 people.[47] If people who read about Willowbrook heard "inevitable" as meaning a likelihood less than 100 percent, then some of the assertions about disease inevitability make more sense. The malleable meanings of the terms may have blunted any desire to interrogate the numerical expressions with the same intensity as the verbal.

Although physicians perceived the disease as mild, having hepatitis clearly involved some element of discomfort. Parents may have thought differently about disease if the inevitability was sooner rather than later. Suffering from hepatitis this year and suffering from hepatitis nineteen years from now are different scenarios with different implications. Although not raised by the investigators, this issue—what we would today call discounting years of healthy life—was well known as early as the eighteenth century.[48] Even if one (or one's child) is among the group (a little more than half) who would have "naturally" become infected with hepatitis, getting hepatitis "this year" is probably worse than getting hepatitis at some unspecified future date. The investigators asserted that it was better to get the disease as a child because it was a less serious disease. However, it does not seem too great a stretch of logic to assume that, were all other things equal (or close), it would be better to become sick at some point in the future than it would be to become sick now.

Discussions and Data. Many social scientists and historians may have been (and continue to be) willing to accept uncritically the numbers and assumptions made by the scientific investigators that they study. In some instances such a stance is due to a perception of their own innumeracy, perhaps based on two misunderstandings. First is the idea that the numbers used in scientific and medical publications are necessarily complicated. At times the numerical analyses used for scientific and medical studies are, indeed, quite complex. While some readers of a complex study may wish to trust the author, others will insist that the author explain her or his methods in terms that permit the reader to understand the study. But the data that we have examined in this paper, data presented in the original Willowbrook studies, hardly qualifies as particularly difficult to comprehend for the college-educated reader, *if* that reader chooses to engage with the question.[49] Second, readers may wish to assume that the numbers have an objective meaning apart from a political context, an objective meaning best ascertained by experts in the field. Social scientists may be naturally drawn to the apparent certainty of numbers while roaming free to critique their creators; perhaps they just "like the sound and appearance of the natural sciences."[50] Moreover, numbers in the context of a medical article may be given more authority than numbers within other contexts, such as numbers in an administrative document. In their book on the political controversies surrounding Willowbrook, David and Sheila Rothman are sensitive to the contingent nature of seemingly objective numbers when presented in a nonscientific context. They interrogate administrative numbers, such as the numbers of nurses needed, in a nuanced and critical fashion that does not allow the authors of the report any right to claim privileged access to "facts" about the number of nurses needed, but the medical claims about probability of disease are accepted as given.[51]

Readers of the medical literature may have a natural tendency to accept the original author's interpretation of the medical and scientific numbers out of regard for the profession. Even while taking a critical view toward some members of that profession, people may wish to respect physicians as representatives of the field of medicine in general. Possible reasons for this attitude are not hard to find. We are all, of course, future patients; and when we are taken ill, we wish deeply to believe that our caregivers will behave in a manner consistent with the prevailing cultural norms of good medicine. We want to believe that physicians know what is "right," at least in a technical sense, and tend to assume that they do.

Seen from another perspective, societal respect for the technical nature of medicine and a belief in the privileged ability of experts to interpret data also contributes to another easily understandable reason that slows critics of numerical data—the fear of being wrong. While disagreements between nonquantitative assumptions may be resolved in ways that allow both sides to be perceived as having a valid point, disagreements about numerical data are often resolved by one side being labeled correct and the other being wrong, and no one wants to wind up on the losing side of such a debate. Hence, historians of human experimentation might tend to avoid a direct debate on the quantitative data unless *very* sure of their approach.

Critics who engaged with primary data on the Willowbrook experiments were often disparaged by the supporters, who claimed that such forays into the experimental world only showed that those who criticized the experiments simply did not understand the science.[52] In some cases this was a valid point, as when people referred incorrectly to the "injection" of live hepatitis virus.[53] While one might wish to argue that it matters little to state (or know) the precise mechanism by which virus was transferred from child to child, there remain significant differences between feeding and injecting, and inaccuracy in one area could breed distrust about the entire argument. Other critics pointed out issues with the quantifiable data.[54] Defenders of the Willowbrook experiments have used technical misunderstandings as evidence for the critics' fundamental lack of understanding of science and thus their lack of standing as people who could intelligently discuss the ethics of the procedures. Thus, critics may have chosen to attack an experiment on grounds other than quantitative analysis, areas where they felt more certain of their abilities.

In addition, policy analysts and historians are trained to look for themes and patterns, and this style of thought may have led them to discount careful numerical analysis. Any criticism of the Willowbrook studies that was based on the specific numerical data, even if a valid critique, would have been perceived as being less interesting, less valuable, because such a criticism would not be easily generalizable to other studies. This set it apart from broader concerns, such as using children for experiments, that could easily be extended to other settings.

One might ask: Why, indeed, should policy analysts and historians care about the primary data—either numerical or not—used by the physicians and scientists that they choose to study? Looking at numbers seems mundane, it seems uninteresting—and sometimes it is. But leaving it to investigators to be the sole arbiter of what their experimental data means can create large problems. By allowing the investigator alone to define the meaning of her or his data, one allows the experimenter and the reader to perpetuate the myth that scientific data is pure, objective, transcultural, and ahistorical. This is a stance that one should not accept. Scholars should not let the object of their study be the privileged reader of the primary experimental data; they should not allow the investigator to be the only one to take the authoritative stance. Obviously, historians cannot spend all of their time re-doing every action, investigation, and computation of the scientists they wish to study. Nor should they. But they should be attuned to the issues and assertions that are most central to the subject of their study and question the basis on which those assertions are being made and accepted. In the case of Willowbrook, a central element of the debate was the assertion that disease was inevitable. The basis for that knowledge claim, and the uses to which that knowledge claim has been put, are worthy of serious attention.

Writing Willowbrook, Reading Willowbrook

In the case of Willowbrook, attention to the inevitability of diseases highlights an ethical issue central to the differences between supporters and critics of medical experiments. Observers from within the medical community, those whose training and professional lives were focused on analysis within the natural sciences, were trained and worked in a world that valued quantitative data but was essentially ahistorical. Knowledge exists; the time course of its creation is not a central topic for scientific inquiry. The Willowbrook experimenters may have had a "hunch" about the incidence of disease before they had the data to back up their clinical intuition. The question of how the knowledge was created may have seemed irrelevant to scientists. Since they (meaning the broader scientific community) were eventually proven right about the incidence of hepatitis at Willowbrook, there was no need to look back at *when* they knew that they were right.[55] But from a historical perspective, what is important is not whether the experimenters eventually knew what the incidence of hepatitis was, but what they knew when they started the experiments.

In addition, the Willowbrook data may have remained essentially unexamined because the narrative story made a lot of sense. Both the critics and the experimenters wanted to see the situation at Willowbrook as being as bad as possible, but for different reasons. For the experimenters, it provided evidence that they couldn't make things any worse. Hepatitis was but one of many diseases than were appallingly prevalent at Willowbrook, and the researchers may have perceived that even aside from the risk of their getting hepatitis, children would likely also get sick from some other sorts of infectious disease. The inevitability of hepatitis provided a wonderful research opportunity, coming at a time when new laboratory techniques were becoming available. For critics, accepting the reading of the inevitability of disease provided evidence that the situation at Willowbrook was intolerable. It was a symbol of the profound social and structural problems that had created Willowbrook. With or without the data, the children were obviously in need of help. Together, the defenders and critics produced a reading of the Willowbrook tables serving the purposes of each.

The many tales of Willowbrook may be instructive for historians of human experimentation and for historians of medicine and science in general. For as the histories are told and re-told from differing disciplinary perspectives, from differing institutional and epistemological stances, the tales may be quite different indeed. Each of the papers in this volume—including this chapter itself—has been written with a particular point of view. By examining the retelling of the Willowbrook experiments, we hope to highlight how any experiment is susceptible to telling and retelling. This is true even for stories that are not as contentious as Willowbrook. Those who wish to examine human experimentation need to attend to how the dominant story has been created and to remain attuned to the possibility that there may be other stories that need attention. In addition, the writing and re-writing of Willowbrook serves to focus on the central issue of how engaged historians and social scientists ought to be in examining the experimental details, as well as the perils of remaining disengaged.

NOTES

Ian Burney, Tracy Crew, Howard Markel, Martin Pernick, Carl Schneider, and Nicholas Steneck read and commented on an earlier draft, and we appreciate their efforts. Members of the Division of General Medicine gave us useful comments at a seminar presentation. Those who attended the February 1995 conference at Columbia College of Physicians and Surgeons, "Regulating Human Experimentation in the United States: The Lessons of History," provided insightful and critical commentary. William Muraskin allowed us to quote from his unpublished paper. Saul Krugman was kind enough to speak with JDH on the telephone.

1. Robert Ward, Saul Krugman, Joan P. Giles, and A. Milton Jacobs, "Endemic Viral Hepatitis at an Institution: Epidemiology and Control," in Frank W. Hartman, Gerald A. LoGrippo, John G. Mateer, and James Barron, eds. *Hepatitis Frontiers* (A Henry Ford Hospital International Symposium) (Boston, 1957), 227–36.

2. We put "mentally retarded" in quotation marks to indicate that it would not be an appropriate term to use today. However, this was the widely used term throughout most of the period, and it is thus the term we shall use to describe the children at the Willowbrook State School.

3. Robert Ward, Saul Krugman, Joan P. Giles, A. Milton Jacobs, and Oscar Bodansky, "Infectious Hepatitis: Studies of Its Natural History and Prevention," *New England Journal of Medicine* 258 (1958): 407–16.

4. John R. Paul, *A History of Poliomyelitis* (New Haven, CT, 1971).

5. David J. Rothman, "Were Tuskegee and Willowbrook 'Studies in Nature'?" *Hastings Center Report* 12 (April 1982): 5–7. Another objection to this logic comes from the fact that the adult caretakers were also subject to exposure and might themselves have made even better experimental subjects because they would have been able to give consent. However, adults might have suffered more serious medical consequences since the disease was generally thought to be more severe in adults than in children. See Paul Ramsey, *The Patient as Person: Explorations in Medical Ethics* (New Haven, CT, 1970).

6. Henry K. Beecher, "Ethics and Clinical Research," *NEJM* 274 (1966): 1354–60. One of the early discussions that is particularly valuable as a primary source is "Proceedings of the Symposium on Ethical Issues in Human Experimentation: The Case of Willowbrook State Hospital Research, May 4, 1972." The symposium was sponsored by the Student Council of New York University School of Medicine. The proceedings were published by the Urban Health Affairs Program, New York University Medical Center (New York, 1972).

7. Krugman offers a spirited defense in the student-run conference (note 6 above), as well as in Saul Krugman, "The Willowbrook Hepatitis Studies Revisited: Ethical Aspects," *Reviews of Infectious Diseases* 8 (1986): 157–62. Also see Daniel S. Gillmor, "How Much for the Patient, How Much for Medical Science? An interview with Dr. Saul Krugman," *Modern Medicine* 30 (7 January 1974): 30–35. In this interview Krugman claims that he warmly supported the idea for the NYU students' symposium. For another hagiographic account with a remarkable opening line, see Allen B. Weisse, "Why Do They Turn Yellow?" *Hospital Practice* (office edition) 26(6) (15 June 1991): 175–76, 178, 183–84, passim. This article opens by noting that "although hepatitis has been recognized as a human affliction for nearly 5,000 years, until only a few decades ago it remained as much a puzzle as plague to the civilized world." The solution, according to the author of the article, came from three people: Krugman, Prince, and Blumberg.

8. Krugman (n. 7 above).

9. Ramsey (n. 5 above).

10. The reaction may also have resulted from a general disinclination to see profoundly retarded children as worthy of much attention (or to see them at all). For discussion of the issues a decade or so later, see A. Sandra Abramson and Constance Bloomfield, "The Politics of Mental Retardation," *Health/PAC Bulletin* no. 48 (January 1973): 1–10. For general background on Willowbrook, see David J. Rothman and Sheila M. Rothman, *The Willowbrook Wars* (New York, 1984) and Ronda Kotelchuck, "Willowbrook: From Agony to Action," *Health/PAC Bulletin* no. 48 (January 1973): 11–18.

11. In a conversation between JDH and Saul Krugman on 13 February 1995, Krugman stated that for answers to such questions, one should consult the printed literature. The other primary investigators were at that time already dead.

12. In writing this work we have made no attempt to look at archival manuscripts to examine how the Willowbrook papers may have changed in the process of being written. The purpose of this research is to look at the debate as it has been discussed in the printed record, using information and publications that would have

been easily available to people working in an academic setting. We have also treated the Willowbrook investigators as a single group led by Saul Krugman. The group obviously did not exist unchanged throughout the period we study. Nonetheless, the work they did is often referred to as a single body of investigation, and we have chosen not to examine the sociology of the research team.

13. Paul B. Beeson, "The Growth of Knowledge about a Disease: Hepatitis," *American Journal of Medicine* 67 (1979):366–70. See also Thomas S. Chen and Peter S. Chen, *Understanding the Liver: A History* (Westport, CT, 1984). This section is also based on the writings of some Willowbrook investigators who frequently reviewed the "state of the art" of knowledge about hepatitis in this period.

14. Robert W. McCollum, "Epidemiological Patterns of Viral Hepatitis," *American Journal of Medicine* 32 (1962): 657–64.

15. Edward B. Grossman, Sloan G. Stewart, and Joseph Stokes, "Post-transfusion Hepatitis in Battle Casualties," *Journal of the American Medical Association* 129 (1945): 991–94.

16. Walter P. Havens, "Period of Infectivity of Patients with Experimentally Induced Infectious Hepatitis," *Journal of Experimental Medicine* 83 (1946): 251–59. Krugman often referred to the importance of the volunteer work that preceded his series of experiments, as in Saul Krugman, "Viral Hepatitis: Overview and Historical Perspectives," *Yale Journal of Biology and Medicine* 49 (July 1976): 199–203; Saul Krugman and Joan P. Giles, "The Natural History of Viral Hepatitis," *Canadian Medical Association Journal* 106 (1972): suppl. 442–46.

17. *Icterus* refers to jaundice, often seen most clearly in the "whites of the eyes," where it is referred to as scleral icterus.

18. Hans Popper and Murray Franklin, "Diagnosis of Hepatitis by Histological and Functional Laboratory Methods," *JAMA* 137 (1948): 230–34; John W. Norcross, Joseph D. Feldman, Robert F. Bradley, and Robert M. White, "Liver Function: An Attempt to Correlate Structural Change with Functional Abnormality," *Annals of Internal Medicine* 35 (1951): 1110–16.

19. For examples and discussion from the group who worked at Willowbrook, see Oscar Bodansky, Saul Krugman, Robert Ward, Morton K. Schwartz, Joan P. Giles, and A. Milton Jacobs, "Infectious Hepatitis: Correlation of Clinical and Laboratory Findings, Including Serum Enzyme Changes," *A.M.A. Journal of Diseases of Children* 98 (1959): 166–86; Joan P. Giles, Saul Krugman, Philip Ziring, A. Milton Jacobs, and Cass Lattimer, "Leucine Aminopeptidase Activity in Infectious Hepatitis," *American Journal of Diseases of Children* 105 (1963): 256–60.

20. Hans Popper, Frederick Steigmann, Karl A. Meyer, Donald D. Kozole, and Murray Franklin, "Correlation of Liver Function and Liver Structure," *American Journal of Medicine* 6 (1949): 278–91; Hans Popper, Frederick Steigman, and Paul B. Szanto, "Quantitative Correlation of Morphological Liver Changes and Clinical Tests," *American Journal of Clinical Pathology* 19 (1949): 710–24.

21. Robert Ward, Saul Krugman, Joan P. Giles, A. Milton Jacobs, and Oscar Bodansky, "Infectious Hepatitis: Studies of Its Natural History and Prevention," *New England Journal of Medicine* 258 (1958): 407–16.

22. In the section devoted to ethical justification, they state that "the annual at-

tack rates of jaundice were high—for example 20 to 25 per 1,000" (p. 412). For later references, see Bodansky et al. (n. 19 above); Krugman (n. 7 above).

23. Ward (n. 21 above); Saul Krugman, Robert Ward, Joan P. Giles, Oscar Bodansky, and A. Milton Jacobs, "Infectious Hepatitis: Detection of Virus During the Incubation Period and in Clinically Inapparent Infection," *New England Journal of Medicine* 261 (1959): 731–34; Saul Krugman, Robert Ward, and Joan P. Giles, "The Natural History of Infectious Hepatitis," *American Journal of Medicine* 32 (1962): 717–28; Krugman (n. 7 above).

24. Ward (n. 21 above). The situation at Willowbrook was complex, with new children entering the system at a varying rate. If these children entered at some point other than the full four-year period on which the data in Table 2 is based, they would have been exposed for less than four years. Yet each would still have been listed as having acquired the disease at an equal risk for each of the four years (including the years before they arrived); this assumption would tend to raise the apparent attack rate. As children continued to enter the system, the overall path of the epidemic would still tend to follow the traditional pattern, but the precise outline of the epidemic is difficult to predict without considerably more data.

25. Krugman (n. 16 above).

26. Already in 1956 it was being noted that some patients appeared to have recurrent hepatitis. The explanation at that time was unclear; we now would assume that the two infections would have been caused by two different viruses, probably hepatitis A and hepatitis B. Factoring this into the calculations would decrease the predicted likelihood of infection, as some of the cases could have represented the same person having hepatitis twice.

27. Saul Krugman and Robert Ward, *Infectious Diseases of Children* (St. Louis, 1958), 79.

28. Robert Ward and Saul Krugman, "Viral Hepatitis," *Disease-a-Month* October 1961.

29. Ward (n. 21 above).

30. Richard B. Capps, Alfred M. Bennett, and Joseph Stokes, "Infectious Hepatitis in an Infants' Orphanage. 1. Epidemiological Studies in Student Nurses," *Archives of Internal Medicine* 89 (1952): 6–23.

31. Krugman is never quite clear on the precise ages of the children. He does suggest in an early report that he is dealing in large part with 5 to 10-year-olds. Saul Krugman, Robert Ward, Joan P. Giles, and A. Milton Jacobs, "Experimental Transmission and Trials of Passive-Active Immunity in Viral Hepatitis," *AMA Journal of Diseases of Children* 94 (1957): 409–11.

32. Ward (n. 1 above).

33. Saul Krugman and Robert Ward, "Clinical Conference: Clinical and Experimental Studies of Infectious Hepatitis," *Pediatrics* 22 (1958): 1016–22.

34. This table was constructed based on the assumptions above by very straightforward use of ratios. If the age-specific four-year attack rate was 51/1,000 for children ages 0–4, then after four years 5.1 percent of the children would be infected. That leaves 949/1,000 children uninfected. For ages 5–9, the age-specific four-year attack rate was 67/1,000. Note that this rate would apply to the children for five

years, from ages 5 through 9 (the earlier rate given in Table 2 was for a four-year period, from ages 0–5). That ratio is then applied to the group of 949 children who would not yet be infected. A similar procedure is applied to arrive at the remainder of the table. The calculations for a 1:1 ratio are obtained by doubling the attack rates. The calculations would not have been difficult, or even remarkable, for people in the 1950s. If one wishes to increase the ratio of asymptomatic to symptomatic cases still more, say, to a 2:1 ratio, one still arrives at an estimate of only 70 percent of the population having hepatitis by the age of 19.

35. We have chosen to stop the analysis at age 19 in part because it seemed logical to look at the period from infancy to adulthood, in part because the investigators themselves used age 19 as a demarcation in their discussion.

36. Saul Krugman, Robert Ward, Joan P. Giles, and A. Milton Jacobs, "Infectious Hepatitis: Studies on the Effect of Gamma Globulin on the Incidence of Inapparent Infection," *JAMA* 174 (1960): 823–30.

37. And, in fact, exceeds it—i.e., if one takes the attack rates as given and assumes a 12:1 ratio, one quickly arrives at estimates of over 100 percent of the population being infected, an impossibility that reflects the fact that neither the attack rates nor the 12:1 ratio are particularly precise estimates. But that does not matter a great deal for purposes of understanding the actions and beliefs of either the investigators or the critics. What is most important is that this ratio gave credence to the claim of "inevitable disease."

38. Saul Krugman, Harriet Friedman, and Cass Lattimer, "Hepatitis A and B: Serological Survey of Various Population Groups," *American Journal of the Medical Sciences* 275 (1978): 249–55.

39. Clearly, after 1978 they felt that the new data about the incidence of hepatitis justified their statements, if not their earlier data. One continuing theme in the retrospective analysis as performed by defenders of the experiments has been that the results justified the experiments. Thus, if they were "right," it probably didn't seem important how, or when, they knew that they were "right."

40. Bodansky et al. (n. 19 above).

41. Ward et al. (note 1 above) discusses the statistical nonsignificance of a finding; other references to significance testing may be found in Ward (note 21 above).

42. Krugman gave this assertion in an interview with William Muraskin, described in "The Willowbrook Experiments Revisited: Saul Krugman and the Politics of Morality," unpublished manuscript, 29 January 1991. We are grateful to Dr. Muraskin for allowing us to read this manuscript and to cite it here, with permission to do so granted in a telephone conversation of 3 January 1995. One should read Krugman's conversations with Muraskin in light of the fact that Muraskin takes a very sympathetic view of Krugman's experiments. Although unpublished, this paper has become part of the public debate, being presented at an extremely contentious session of the 1990 annual meeting of the American Association for the History of Medicine.

43. Ward (n. 21 above).

44. Geoffrey Edsall, "Experiments at Willowbrook," *Lancet* 2 (1971): 95.

45. Edward N. Willey, "Experiments at Willowbrook," *Lancet* 2 (1971): 1078.

46. One little-noted exception is Louis Goldman, *When Doctors Disagree: Con-*

troversies in Medicine (London, 1973), 71: "Certainly viral hepatitis was very common at Willowbrook, but the annual attack rate was put only at 20 to 25 per 1,000. Admittedly, there were probably many more sub-clinical, undiagnosed cases; no one can be absolutely sure how many this would be. Yet Dr. Krugman . . . claimed in his 1970 paper that infection was 'inevitable.' Surely 'inevitable' means 100 per cent. No figures seem to have been produced to show that *all* of the children at Willowbrook develop hepatitis at one time or another, sooner or later." Muraskin in "The Willowbrook Experiments Revisited" also addresses this point, but he accepts Krugman's explanation.

47. Kimberly Koons Woloshin, Mack T. Ruffin IV, and Daniel W. Gorenflo, "Patients' Interpretation of Qualitative Probability Statements," *Archives of Family Medicine* 3 (1994): 961–66. See also D. J. Mazur and D. H. Hickam, "Patient Interpretations of Terms Connoting Low Probabilities when Communicating About Surgical Risk," *Theoretical Surgery* 8 (1993): 143–45; Geoffrey D. Bryant and Geoffrey R. Norman, "Expressions of Probability: Words and Numbers," *New England Journal of Medicine* 302 (1980): 411; Robert M. Kenney, "Between Never and Always," *New England Journal of Medicine* 305 (1981): 1097–98.

48. See the discussion over smallpox inoculation between Daniel Bernoulli and Jean D'Alembert as described in Lorraine Daston, *Classical Probability in the Enlightenment* (Princeton, NJ, 1988), 83–89.

49. Perhaps the most confusing part of the way that the data is presented is that rather than listing attack rates per year, the rates are presented for a four-year period, a decision noted in a footnote.

50. Charles Leslie, "Scientific Racism: Reflections on Peer Review, Science, and Ideology," *Social Science Medicine* 31 (1990): 891–912, 904.

51. Rothman and Rothman (n. 10 above).

52. See, for example, Gilmore (n. 7 above). Saul Krugman, in his telephone interview with JDH commented that David and Sheila Rothman, the two authors of the major book-length interpretation of Willowbrook, both nonphysicians, "simply did not understand hepatitis," although he did not elaborate.

53. David S. Shimm and Roy G. Spece, "Conflict of Interest and Informed Consent in Industry-Sponsored Clinical Trials," *Journal of Legal Medicine* 12 (1991): 477–513, 479.

54. Henry K. Beecher, *Research and the Individual: Human Studies* (Boston, 1970).

55. For example, M. F. Perutz defends Louis Pasteur against what he sees as an overly critical biography in large part because Pasteur was, after all, "right." M. F. Perutz, "The Pioneer Defended," review of Gerald L. Geison, *The Private Science of Louis Pasteur* (Princeton, NJ, 1995), in *New York Review of Books*, 21 December 1995, 54–58.

Contributors

Brian Balmer is a senior lecturer in science policy studies in the Department of Science and Technology Studies, University College London. He has published widely on the sociology and politics of research in biotechnology and human genetics. His current research is on the history of biological warfare policy. He is the author of *Britain and Biological Warfare: Expert Advice and Science Policy, 1935–1965*.

Miriam Boleyn-Fitzgerald holds a degree in physics from Swarthmore College and has worked on the staffs of the White House Advisory Committee on Human Radiation Experiments, the Union of Concerned Scientists, and the National Resources Defense Council. She is a writer and Developmental Math Specialist at the University of Wisconsin–Fox Valley.

Jordan Goodman is a historian of science and medicine recently working at the University of Manchester Institute of Science and Technology and currently Honorary Research Fellow at the Wellcome Trust Centre for the History of Medicine and the Department of Anatomy and Developmental Biology at University College London. He has published widely in the history of medicine, science, and technology, including several articles on the history of therapeutics and a recent book, *The Story of Taxol: Nature and Politics in Pursuit of an Anti-Cancer Drug*. He is completing a book on a scientific voyage in the mid-nineteenth century.

Rod Hayward is a general internist who received his training in biostatistics and epidemiological methods as a Robert Wood Johnson Clinical Scholar at UCLA and at the RAND Corporation, Santa Monica. He is currently a Professor of Medicine and Public Health at the University of Michigan and the Director of the VA Center for Health Services Research and Development in Ann Arbor, Michigan. His research focuses on statistical and methodological aspects of quality measurement and improvement for chronic illnesses.

Joel Howell is the Victor Vaughan Professor of the History of Medicine at the University of Michigan. He is a clinically active internist, as well as Professor of History and of Health Management and Policy. He has published widely on the history of medicine, including *Technology in the Hospital: Transforming Patient Care in the Early Twentieth Century.*

Margaret Humphreys is the Josiah Charles Trent Associate Professor of Medical Humanities at Duke University, where she teaches the history of medicine and public health. Her most recent book is *Malaria: Poverty, Race, and Public Health in the United States.* She is currently researching projects on the American Civil War and racial issues in medicine.

David Jones studied medicine and history of science at Harvard University. His forthcoming book, *Rationalizing Epidemics,* examines responses to American Indian epidemics since the colonial period. He has also published several articles about clinical research after World War II, including the development of coronary artery bypass surgery and the history of outpatient treatment programs for tuberculosis on the Navajo reservation. He is currently a resident in psychiatry at McLean Hospital and Massachusetts General Hospital.

Lara Marks is a visiting senior research associate at Cambridge University and an Honorary Senior Lecturer at the London School of Hygiene and Tropical Medicine. She published *Sexual Chemistry: A History of the Pill.* She has also published numerous books and articles on the history of maternal and child health and ethnicity and health. Her present research concerns the process of drug discovery and development and the application of information technology in the pharmaceutical industry.

Anthony McElligott is the founding Professor of History at the University of Limerick where he is also director of the Centre for Historical Research. His teaching and research interests are in the social, cultural, and political history of Europe and Germany in the twentieth century. His most recent books are *Opposing Fascism: Community, Author-*

ity, and Resistance in Europe (with Tim Kirk) and *The German Urban Experience, 1900–1945: Modernity and Crisis.* He is currently writing a monograph on murder and forensic science in interwar Germany. He was elected a fellow of the Royal Historical Society in 1999.

Robert L. Martensen is the Knight Professor of Ethics and Humanities and Professor of Surgery at Tulane University School of Medicine and Senior Fellow in Tulane's Center for Ethics and Public Affairs. He has published papers on the history of medical physics and the history of bioethics among other topics. Currently, he is completing a book on *The Cerebral Body: Origins and Cultural Politics.*

Glenn Mitchell is senior lecturer in the History and Politics Program, University of Wollongong. His published works include *On Strong Foundations: The BWIU and Industrial Relations in the Australian Construction Industry, 1942–1992;* health policies for ethnic minorities for state and federal governments; and essays on Australian housing, environmental pollution, and oral history.

Jenny Stanton is an honorary lecturer at the London School of Hygiene and Tropical Medicine and has just edited *Innovations in Health and Medicine: Diffusion and Resistance in the Twentieth Century.* Her recent research has been on high-technology medicine.

Gilbert Whittemore is a historian of science and an attorney based in Boston with a general interest in the interaction of law and science. He served on the staff of the 1994–1995 Advisory Committee on Human Radiation Experiments and is currently writing a history of radiation safety standards in the United States. His most recent publication is "The Multidimensional Chess of Science and Society: A Postwar Debate over Plutonium Exposure" in *Science, Social History, and Activism,* ed. Garland Allen and Roy MacLeod.